POLYOMINOES

SOLOMON W. GOLOMB

Polyominoes

PUZZLES,
PATTERNS,
PROBLEMS,
AND
PACKINGS

SECOND EDITION

WITH MORE THAN 190 DIAGRAMS
BY WARREN LUSHBAUGH

PRINCETON UNIVERSITY PRESS
PRINCETON, NEW JERSEY

Library of Congress Cataloging-in-Publication Data

Golomb, Solomon (Solomon Wolf)
Polyominoes: Puzzles, Patterns, Problems, and Patterns/
    Solomon W. Golomb : with more than 190 diagrams.
        p.   cm.
Includes bibliographical references and index.
ISBN 0-691-08573-0
1. Polyominoes.
QA166.75.G65   1994
2d. ed.
511'.6—dc20   93-41756

This book has been composed in Times Roman and Helvetica

For Astrid and Beatrice

Contents

Preface to the Revised Edition

THE ORIGINAL edition of *Polyominoes* appeared in 1965, when the subject was still fairly new, and I could include all the material then available that I considered genuinely interesting. Some three decades later the situation is certainly different. A definitive treatment of all that is known, or even of everything reasonably interesting that has been discovered, would fill many volumes. My hardest task in producing this revised edition was deciding what new material to include and what to omit. I believe I have succeeded rather well. I hope you will agree.

Ever since the original edition went out of print, many people have asked me how they could obtain a copy, or, at the very least, when it would be reissued. With that in mind, I have preserved the original text as much as possible, correcting only a few minor errors, mostly typographical, from that edition. However, I have added two new chapters (8 and 9), which cover what I consider to be the most interesting and important subsequent developments. The Problem Compendium, which appears again in this edition, concluded with twelve unsolved problems recommended for "readers' research." Professor Andy Liu of the University of Alberta, in Edmonton, Canada, has kindly consented to have his recent article on the current status of these problems included as Appendix C in this edition. I have also added an update on the status of the enumeration of both plane and solid n-ominoes (Appendix D), entitled "Klarner's Konstant and the Enumeration of N-ominoes." Finally, there is a greatly expanded bibliography, covering much of the vast literature on polyominoes that has developed over the years.

It is impossible to acknowledge, or even to identify, all those whose efforts have enriched this subject. Many people are mentioned in the text when their specific contributions are described, and others appear as authors of referenced works. However, there are several who I feel deserve special acknowledgment here.

Martin Gardner has been an inspiration to everyone writing about recreational aspects of mathematics. His long-running *Scientific American* column was what first brought polyominoes to the attention of a wide audience. He has served as a central clearinghouse for new ideas and results found by the readers of his columns, and he provided me with a major source of bibliographic information for this edition.

David A. Klarner has been a leading researcher and contributor to the study of polyominoes for more than thirty years. He has been extraordinarily helpful in the compilation of both the original edition and this new edition of *Polyominoes*.

Joseph S. Madachy, editor of the *Journal of Recreational Mathematics* and the earlier *Recreational Mathematics Magazine*, has consistently provided a publishing outlet for articles and problems on polyominoes by many contributors.

Basil Gordon and Bruce Rothschild, editors of the *Journal of Combinatorial Theory*, have accepted and published a sizable fraction of the important articles that treat polyominoes as an area of serious mathematical research.

Aviezri Fraenkel of Israel, Richard Guy of Canada, David Singmaster of England, and John Selfridge of the United States contributed information for the bibliography and have been helpful in many other ways as well.

Others whose interest in polyominoes specifically, and in mathematical recreations generally, deserves special mention include Elwyn R. Berlekamp, John H. Conway, and Herbert Taylor. Notably, I am indebted to Donald E. Knuth for supplying a wealth of information and references.

Finally, I wish to thank Princeton University Press for their interest in publishing this new edition, and the great army of polyomino fans who have made this subject so popular.

Los Angeles, California
March 1994

Preface to the
First Edition

EVER SINCE I "invented" polyominoes in 1953 in a talk to the Harvard Mathematics Club, I have found myself irrevocably committed to their care and feeding. A steady stream of correspondents from around the world and from every stratum of society—board chairmen of leading universities, residents of obscure monasteries, inmates of prominent penitentiaries— have asked for further information, posed new problems, or furnished novel solutions.

With the passage of time, I learned of the true antiquity of pentominoes, one kind of polyomino. Although the name was coined in my lecture of 1953, the first pentomino *problem* was published in 1907 in the *Canterbury Puzzles*, written by the great English inventor of puzzles, Henry Ernest Dudeney; and the observation that there are twelve distinctive patterns (the pentominoes) that can be formed by five connected stones on a Go board (the very old Japanese game is played with black and white stone markers placed on a board) is attributed to an ancient master of that game. Moreover, an extensive literature on the subject (under the heading of "dissection problems" rather than "polyominoes") had appeared during the 1930s and 1940s in the *Fairy Chess Review*, a British puzzle journal.

A year after it had been delivered, my Harvard talk was published in *American Mathematical Monthly*, where it attracted the attention of a number of professional mathematicians. However, it was the reprinting of some of this material in the May 1957 issue of *Scientific American* that brought polyominoes to the attention of a vast reading public.

Since 1957, many groups of high school and college students and teachers have requested lectures on polyominoes, and as well as delivering many talks about this mathematical recreation, I have written numerous articles about it. Polyominoes not only have wide appeal among mathematical-recreations fans, they also serve as fascinating "enrichment" material in school mathematics programs. The lectures, articles, and voluminous correspondence have all helped to precipitate the present book.

Grateful appreciation is expressed to the former publication, *Recreational Mathematics Magazine*, and its editor, Joseph S. Madachy, for permission to incorporate material that first ap-

Preface to
First Edition

peared in that publication. The author is also grateful to *Scientific American* for permission to use material previously published in Martin Gardner's "Mathematical Games" column.

The tolerance of the Jet Propulsion Laboratory and its director, Dr. W. H. Pickering, of my extracurricular interest in polyominoes during a long association with that organization is deeply appreciated. The meticulous typing and manuscript preparation of Julie Jacobs deserves special commendation. Invaluable assistance in editing, proofreading, figure preparation, and problem selection—including the Problem Compendium at the end of this book—was supplied by Warren Lushbaugh, whose active participation accelerated the completion of the manuscript by many months. Finally, the contributions of a vast army of polyomino fans cannot go unnoticed; special mention must be made of David Klarner and Spencer Earnshaw, who removed many of the most difficult problems from the "unsolved" category.

SOLOMON W. GOLOMB

University of Southern California,
Los Angeles, 1964

POLYOMINOES

Polyominoes and Checkerboards

1

THIS BOOK explores polyominoes, shapes made by connecting certain numbers of equal-sized squares, each joined together with at least one other square along an edge. (Chess players might call this "rookwise connection"; that is, a rook—which can travel either horizontally or vertically in any one move, but never diagonally—placed on any square of the polyomino must be able to travel to any other in a finite number of moves.)

Polyomino patterns are actually examples of *combinatorial geometry*, that branch of mathematics dealing with the ways in which geometrical shapes can be combined. It is a frequently neglected aspect of mathematics because it seems to have few general methods, and because in it systematic rules have not replaced ingenuity as the key to discovery. Many of the design problems in practical engineering are combinatorial in nature, especially when standard components or shapes are to be fitted together in some optimal fashion. The aim of this chapter is twofold: first, to serve as an introduction to the mathematical recreation of polyominoes; and second, to illustrate some of the thinking that can be used effectively whenever problems in combinatorial geometry arise.

The simpler polyominoes—all the possible shapes composed of fewer than 5 connected squares—are shown in figure 1. In the combinatorial problems that follow it will be assumed that polyominoes can be *rotated* (turned 90, 180, or 270 degrees) or *reflected* (flipped over) at will unless otherwise specified (see the section on one-sided polyominoes, page 70).

Monomino

Domino

Straight Tromino

Right Tromino

Straight Tetromino

Square Tetromino

T Tetromino

Skew Tetromino

L Tetromino

Figure 1. The simpler polyomino shapes.

DOMINOES

A *domino* is made of 2 connected squares and has only one shape, a rectangle. The first problem, with which some readers of this book may be familiar, concerns dominoes: Given a checkerboard with a pair of diagonally opposite corner squares deleted (see figure 2) and a box of dominoes, each of which covers exactly 2 squares, is it possible to cover this board completely with dominoes (allowing no vacant squares and no overlaps)?

Figure 2. Checkerboard with opposite corners deleted.

The answer is no, and a remarkable proof can be given. The standard checkerboard contains 64 squares of alternating light and dark colors (referred to hereafter as *checkerboard coloring*). On this board, each domino will cover one light square and one dark square. Thus, *n* dominoes (any specific number of dominoes) will cover *n* light squares and *n* dark squares, that is, a number of each equal to the total number of dominoes. However, the defective checkerboard has more dark squares than light ones, so it cannot be covered. This result is really a theorem in combinatorial geometry.

TROMINOES

It is impossible to cover an 8 × 8 board entirely with *trominoes*, polyominoes of 3 squares, because 64 is not divisible by 3. Instead, it shall be asked: Can the 8 × 8 board be covered with 21 trominoes and one *monomino* (a single square)?

First, suppose 21 straight trominoes are used; the board is colored "patriotically" (see figure 3), and it is observed that a straight tromino will cover one red square, one white square, and one blue square, no matter where the piece is placed. Thus, 21 straight trominoes will cover 21 each of the red, white, and blue squares. By actual count, 22 red, 21 white, and 21 blue squares are involved in the 3-colored 8 × 8 board.

Key:

Red ▨

White □

Blue ■

Figure 3. The 3-colored checkerboard.

If a monomino is placed in the lower left-hand square, the remaining board will consist of 22 red, 21 white, and 20 blue squares. Thus, the board cannot be covered with 21 straight trominoes and a monomino in the lower left-hand corner. If some other corner had been covered with the monomino, the board could have been rotated until the monomino was at the lower left, and it would then have been possible to proceed as before. All four corners, thus, are *symmetric* to one another. That is, the board can be moved by rotation and by reflection in such a way that any corner can be interchanged with any other one. If a construction is impossible in one situation, it remains impossible in any other situation that is symmetric to the first.

Symmetry arguments are very powerful tools in combinatorial geometry. For example, by this reasoning it can be determined that if a monomino is placed on any blue square, or on any white square, or on any square symmetric to a blue or a white, the rest of the board cannot be covered with straight trominoes.

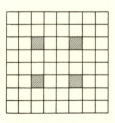

Figure 4. Red squares that are not symmetric to any white or blue squares in the 3-colored checkerboard.

The only red squares not symmetric to blue or white ones are the 4 shown in figure 4. It already has been proved that if a monomino is placed anywhere *except* on one of these four squares, the rest of the board cannot be covered with straight trominoes. The symmetry principle suggests that these four remaining ones *might* be exceptional. The construction of figure 5 shows that they actually are. It is possible to cover the checkerboard with 21 straight trominoes and one monomino, provided that the monomino is placed on one of the four exceptional squares.

When another type of tromino is considered, the result is surprisingly different: No matter where on the checkerboard a monomino is placed, the remaining squares always can be covered with 21 right trominoes.

Consider first a 2 × 2 board. Wherever a monomino is placed, the other three squares can be covered by a right tromino (see fig. 6). Next consider a 4 × 4 board. Divide it into quarters, each of which is a 2 × 2 board. Let the monomino be placed in one of the quarters, say the upper left. The rest of this section can be covered with a right tromino, since it is a 2 × 2 board. In each of the other three quadrants, if a single square is removed, the remaining squares can be covered with a right tromino. And a right tromino placed in the center of the board removes one square from each of three quadrants, making it possible to complete the covering using only right trominoes.

The 8 × 8 checkerboard is treated in the same way. First, divide it into quadrants, each of which will be 4 × 4. The monomino must be in one of the four sections, each of which can be completed because it is a 4 × 4 board. The other quadrants can be covered if one square is removed from each, for this would make them equivalent to the 4 × 4 board with the monomino in it. And these three extra squares can be juxtaposed to form a right tromino in the center of the board.

The proof just given proceeds by *mathematical induction*, a method of formal mathematical proof. The first board was 2 × 2; this could also be written as $2^1 \times 2^1$. (The number 2 is called the *base*, and the superscript 1 is the *exponent*, indicating the *power* to which the base should be raised or the number of times it should be multiplied by itself. Thus, 2^1, or 2 to the first power, is simply 2; 2^2 is 2 × 2, or 4; and 2^3 is 2 × 2 × 2, or 8; 2^n is 2 to the *n*-th power. The 2 × 2 board could also be written as $2^n \times 2^n$, when *n* is equal to 1.) The 2 × 2 case ($2^n \times 2^n$, $n = 1$) was very easy, and the $2^{n+1} \times 2^{n+1}$ case (for example, $2^{1+1} \times 2^{1+1} = 2^2 \times 2^2 = 4 \times 4$) follows readily from the $2^n \times 2^n$ case. Such proofs are very valuable in combinatorial analysis. They suggest that complex geometrical patterns can be achieved by the systematic repetition and combination of simple patterns.

Polyominoes and Checkerboards

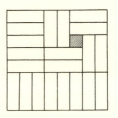

*Figure 5.
A checkerboard covered by 21 straight trominoes and 1 monomino.*

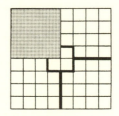

*Figure 6.
Progressive covering by right trominoes.*

TETROMINOES

Some theorems about *tetrominoes* (polyominoes of four squares) are worth mentioning, although detailed proofs will be omitted. Accordingly, each of the following statements may be regarded as a problem exercise.

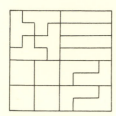

Figure 7. Any tetrominoes except the skew tetrominoes can be used to cover the checkerboard.

1. It is easy to cover the checkerboard entirely with straight tetrominoes, square tetrominoes, T tetrominoes, or L tetrominoes. (This is clear from figure 7.)
2. It is impossible to cover the board, or even a single edge of it, with skew tetrominoes.
3. It is impossible to cover the checkerboard with 15 T tetrominoes and one square tetromino. (This can be proved using the ordinary coloring of the board. One must keep track of even, as opposed to odd, numbers of squares covered.)
4. It is likewise impossible to cover the 8 × 8 board with 15 L tetrominoes and one square tetromino. (Now, however, the most convenient proof uses the dotted and blank squares of figure 8.)
5. It is also impossible to cover the board with one square tetromino and any combination of straight and skew tetrominoes. (The proof in this case makes use of the wavy-lined and uncolored squares arranged as shown in figure 9.)

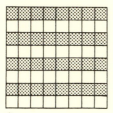

Figure 8. The coloring of the 8 × 8 board used to study coverings by L tetrominoes.

PENTOMINOES

The shapes that cover five connected squares are called *pentominoes*. There are twelve of these and the letter "names" in figure 10 are recommended for them. As a mnemonic device, one has only to remember the end of the alphabet (TUVWXYZ) and the word FILiPiNo. Since there are twelve distinct pentomino shapes, each covering five squares, their total area is sixty squares.

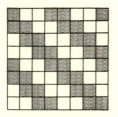

Figure 9. The coloring used to study coverings by straight and skew tetrominoes.

There are numerous ways to place all twelve distinct pentominoes on an 8 × 8 board, with four squares always left over. Many interesting patterns can be formed by artistically specifying the positions of the four extra squares. Three of these patterns are illustrated in figure 11.

Another obvious possibility is to require that the four surplus squares form a 2 × 2 area (a square tetromino) in some specified position on the board. (Two favorite locations are the center and one of the corners.) This placement results in a very remarkable theorem, which can be proved using only

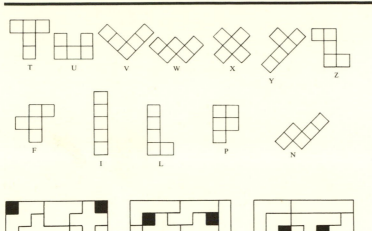

F

I

L

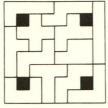

P

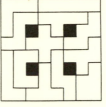

N

*Figure 10. The 12
pentominoes.*

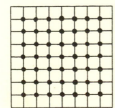

*Figure 11. Three
patterns with all
twelve distinct pent-
ominoes on a single
checkerboard.*

three constructions: Wherever on the checkerboard a square
tetromino is placed, the rest of the board can be covered with
the twelve pentominoes.

At first glance, there are forty-nine possible locations for
the square tetromino, and the heavy dots in figure 12 designate
these positions for the center of the 2 × 2 square. However,
when symmetry principles are applied, the problem reduces
to the ten nonsymmetric positions indicated by the dots in figure 13.

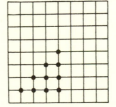

*Figure 12. The
49 possible positions
for the center of a
square tetromino on
the checkerboard.*

*Figure 13. The
10 nonsymmetric
positions for a
square tetromino on
the checkerboard.*

A clever stratagem is to combine the square tetromino with
the V pentomino to form a 3 × 3 square, as shown in figure
14.

*Figure 14. The combination of a square tetromino
and a V pentomino into a 3 × 3 square.*

Then, the three diagrams of figure 15 show the complete proof, because any of the ten positions for the square tetromino can be realized by first selecting the correct diagram and then utilizing the proper position for the 2 × 2 square within the 3 × 3 square.

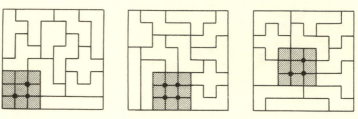

Figure 15. Three constructions prove that anywhere a 2 × 2 square is removed from the checkerboard, the remaining 60 squares can be covered by the 12 distinct pentominoes.

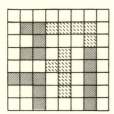

Figure 16. Five pentominoes span the checkerboard.

It is also natural to inquire: What is the *least* number of pentominoes that will span the checkerboard? That is, *some* of the pentominoes are placed on the board in such a way that none of the remaining ones can be added. The minimum number needed to span the board is five, and one such configuration is shown in figure 16.

Many other patterns can be formed using the twelve pentominoes, and the reader may wish to try some of them. Such configurations include rectangles of 6 × 10, 5 × 12, 4 × 15, and 3 × 20. The most difficult of these rectangles is the 3 × 20, and the solution given in figure 17 is known to be unique, except for the possibility of rotating the shaded central portion by 180 degrees.

Figure 17. The 12 pentominoes form a 3 × 20 rectangle.

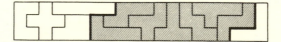

R. M. Robinson, professor of mathematics at the University of California at Berkeley, has proposed another fascinating construction with pentominoes, which he calls the "triplication problem": Given a pentomino, use nine of the other pentominoes to construct a scale model, three times as wide and three times as high as the given piece. Solutions are shown in figure 18 for the V and X pentominoes. The reader is invited to try to triplicate the other ten pentominoes; all of these constructions are known to be possible.

Besides its fascination as a puzzle, the placement of pentominoes on the checkerboard can also make an exciting competitive game. Two or more persons play with one set of the

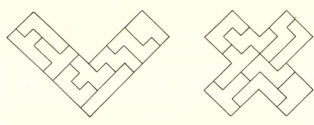

Figure 18. Triplication of the V pentomino and the X pentomino.

twelve pieces, each player placing a pentomino of his choice on an initially empty checkerboard. The first person who is unable to find room on the board for any of the unused pentominoes is the loser. If all twelve pentominoes are successfully placed on the board, the player who placed the last piece is the winner. The game will last at least five, and at most twelve, moves, can never result in a draw, has more possible openings than chess, and will intrigue players of all ages. It is difficult to advise what strategy should be followed, but there are two valuable strategic principles:

1. Try to move in such a way that there will be room for an *even number* of pieces. (This applies only when there are two players.)
2. If a player cannot analyze the situation, he should do something to complicate the placement so that the next player will have even more difficulty analyzing it than he did.

The following is a pentomino problem of a rather different nature from those that have been discussed: A man wishes to construct the twelve pentominoes out of plywood. His saw will not cut around corners. What is the smallest plywood rectangle from which he can cut all twelve pentominoes? (The U pentomino, shaded in figure 19, will require special effort. Assume that it must be cut as a 2 × 3 *hexomino*, a polyomino of six squares, and finished later.) The best answer is not known, but a 6 × 13 rectangle may be used. In the illustration (fig. 19), the heavier lines are to be cut first, starting from the sides and working inward.

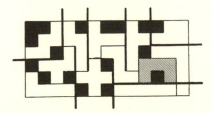

Figure 19. The 12 pentominoes to be cut from a 6 × 13 rectangle.

There is a lesson in plausible reasoning to be learned from
the pentominoes. Given certain basic data, one labors long and
hard to fit them into a pattern. Having succeeded, one then
believes the pattern to be the *only* one that "fits the facts,"
indeed, that the data are merely manifestations of the beau-
tiful, comprehensive whole constructed from them. The pentominoes illustrate that many different patterns may be constructed from the same data, all equally valid, and that the
nature of the final pattern is determined more by the desired
shape than by the information at hand. It is also possible that,
for certain data, no pattern of the type the constructor is con-
ditioned to seek may exist. This will be illustrated by the hexominoes.

HEXOMINOES

There are thirty-five distinct hexominoes and 108 distinct
heptominoes (polyominoes of seven squares). No one has yet
succeeded in obtaining an expression or formula for the exact
number of *n*-ominoes as a function of *n*; that is, a formula that
will give the number of differently shaped polyominoes for
any specified number of connected squares. Combinatorial
problems of this sort are often tantalizingly difficult. Partial
results of calculations on this "polyomino enumeration problem" will be taken up in chapter 6.

The thirty-five hexominoes cover a total area of 210 squares.
It is natural to try to arrange them in rectangles, either 3 ×
70, 5 × 42, 6 × 35, 7 × 30, 10 × 21, or 14 × 15. All such
attempts, however, will fail. To prove this, in each of the
rectangles checkerboard coloring could be introduced, resulting in 105 light and 105 dark squares, an *odd* number of each.
There are twenty-four hexominoes that always will cover three
dark squares and three light squares (an odd number of each).
But the other eleven hexominoes always cover two squares of
one color and four of the other, an *even* number of each. The
thirty-five hexominoes are shown in figure 20 according to
their checkerboard-covering characteristics.

There are an even number of "odd" hexominoes and an odd
number of "even" hexominoes. Since in all multiplication "even
times odd equals even" and "odd times even equals even,"
the thirty-five hexominoes always will cover an even number
of light squares and an even number of dark squares. However, the number of light (or dark) squares is 105 for any of
the rectangles in question, and 105 is odd, so the construction
is impossible.

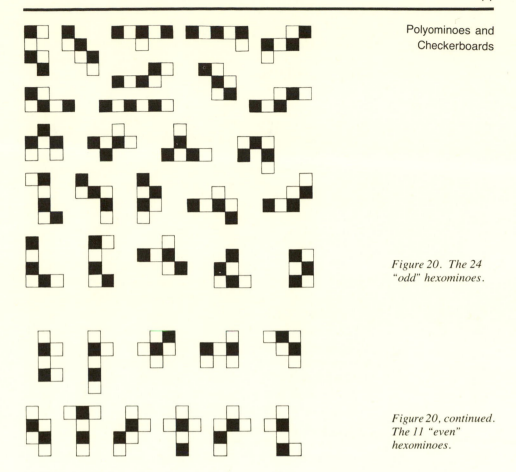

*Figure 20. The 24
"odd" hexominoes.*

*Figure 20, continued.
The 11 "even"
hexominoes.*

It is noteworthy that the same checkerboard coloring—that is, the alternation of light and dark squares—used to prove the simplest fact about dominoes also serves to prove a far more complex theorem about hexominoes. The underlying theme of this coloring is *parity check*, or a check for evenness, a simple, yet powerful, mathematical tool based on the obvious fact that an even number is never equal to an odd number. The use of "colors" is a valuable aid to the intuition—objects colored differently will seldom be confused. And sometimes, as in the straight-tromino problem, the colors vividly proclaim a solution that might otherwise have been overlooked.

2 Patterns and Polyominoes

CHAPTER 1 introduced polyominoes and presented problems that use these shapes in attempts to cover squares and rectangles of various sizes. The present chapter is devoted to a variety of problems that involve fitting these polyominoes together into other specified shapes and patterns.

PENTOMINO PATTERNS

A new class of pentomino patterns to be discussed here are the *superposition* problems, the construction of two or more shapes all of whose parts coincide. Several examples will now be considered.

1. The reader is challenged to arrange the twelve pentominoes into *two* 5 × 6 rectangles of six pentominoes each; the two sets are drawn in figure 21. The choice of sets shown is unique. In the solution of the rectangle made from the set on the right, the F and N pentominoes can be fitted together in another distinct way and still occupy the same region. Note that the answer to this superposition problem simultaneously solves the 5 × 12 and the 6 × 10 rectangle problems simply by putting the two 5 × 6 rectangles together in two different ways.

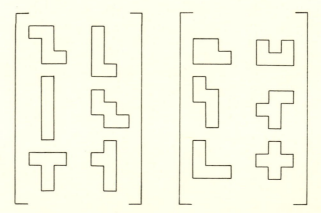

Figure 21. The 2 sets of 6 pentominoes from which a pair of 5 × 6 rectangles can be constructed.

2. Find solutions to the 8 × 8 pattern with the 4-square hole in the middle so that the pieces will separate into two congruent parts, each using six of the pentominoes. Three typical arrangements are shown in figure 22.

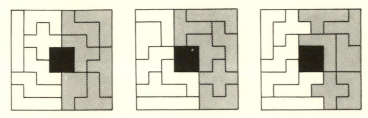

Figure 22. Typical solutions to the 8 × 8 board with a 2 × 2 hole and 2 congruent pieces.

3. Divide the twelve pentominoes into three groups of four each. Find one 20-square region that *each* of the three groups will cover. One solution is shown in figure 23, but several other answers to this problem have been discovered, and readers are invited to look for their own.

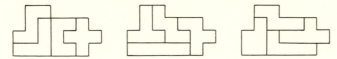

Figure 23. A solution to the "3 congruent groups" problem.

4. Divide the twelve pentominoes again into three groups of four each. Subdivide each group into two pairs of pentominoes. For each group, find a 10-square region that both of its pairs will cover. One solution is given in figure 24. It is interesting to find other configurations, especially those eliminating "holes" in all three regions. The reader may wish to look for a different solution with holes and for one without holes; these do exist.

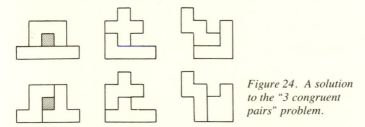

Figure 24. A solution to the "3 congruent pairs" problem.

5. Once more, divide the twelve pentominoes into three groups of four each. To each group add a monomino and form a 3 × 7 rectangle (see fig. 25). This solution is known to be

unique, except that in the first rectangle the monomino and the Y pentomino can be rearranged and still occupy the same region.

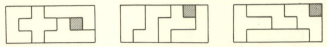

Figure 25. Solution to the three 3 × 7 rectangles of pentominoes.

Figure 26. The only location for the X pentomino in a 3 × 7 rectangle.

The uniqueness proof follows a suggestion of Dr. C. S. Lorens, an engineer employed by the Aerospace Corporation in Los Angeles. To begin with, the X pentomino can be used only in conjunction with the U pentomino in the pattern shown in figure 26. Next, neither the F nor the W pentomino can be used to complete this rectangle. Also, with the U pentomino needed to support the X, it is impossible to use both the F and W in the same 3 × 7 rectangle. Hence, of the three 3 × 7 rectangles, one will contain X, U, another will contain W (but not U), and the third will contain F (but not U). When all possible completions of these three rectangles are listed and compared (a very time-consuming enterprise), it is found that the only possible solution is given in figure 25.

6. Divide the twelve pentominoes into four groups of three each. Find a 15-square region that *each* of the four groups will cover exactly. No solution to this problem is known. On the other hand, the configuration has not been proved impossible.

7. Find the smallest region on the checkerboard onto which each of the twelve pentominoes will fit, one at a time. The minimum area is nine squares, and the two regions are shown in figure 27.

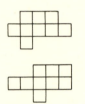

Figure 27. Minimal regions on which each of the 12 pentominoes can fit.

The adequacy of these minimal areas is proved by observing that each pentomino, when placed in the area, does fit. The impossibility of the adequacy of *fewer* than nine squares is shown as follows. If it were possible to use an area of fewer than nine squares, then the I, X, and V pentominoes would fit on a region of no more than eight squares. The I and X pentominoes would then have three squares in common. (Otherwise, either nine squares are needed for the I and X, or else the center square of the X coincides with an end square of the I, and nine squares would be required as soon as the V pentomino is considered.) This can be constructed in two distinct ways, as shown in figure 28. In either case, the position of the V pentomino is then specified. However, the placing of the U pentomino would then require one more square. Thus, eight squares are not enough, whereas nine have been shown by example to be sufficient.

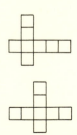

Figure 28. Minimum regions for the I, X, and V pentominoes.

Several years ago, the resources of modern electronic computing were devoted to various pentomino problems. A technical report (for bibliographic information on all material cited in the text, see the Bibliography) by the American logician Dana S. Scott, professor of philosophy at Stanford University in California, describes two constructions that were solved by the MANIAC computer. The first problem was how to fit the twelve pentominoes onto a 3 × 20 rectangle. It was verified that the two solutions already known are indeed the only possible ones. The second problem was to find all the ways to fit the twelve pentominoes onto the 8 × 8 board, leaving a 2 × 2 hole in the center. It was discovered that there are sixty-five basically different solutions (in the sense that two solutions differing only by rotation or reflection of the board are not regarded as distinct). The program included the astute observation that there are only three basically different locations for the X pentomino, as shown in figure 29. It is possible to finish covering the board in each of these cases in 20, 19, and 26 different ways, respectively. Three of the more interesting solutions are shown in figure 30. A number of otherwise plausible situations turn out to be impossible, since they are absent from Scott's listing. These include the configurations shown in figure 31.

Professor C. B. Haselgrove of Manchester University in England, who is mostly known for his contributions to number theory, had also programmed a computer to find all the ways

Patterns and Polyominoes

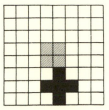

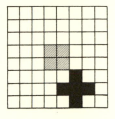

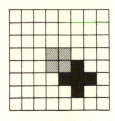

Figure 29. The 3 possible locations for the X pentomino on the 8 × 8 board with a 2 × 2 center hole.

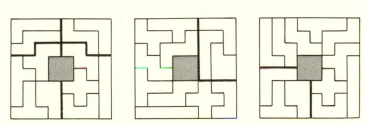

Figure 30. Three interesting solutions to the 8 × 8 board with a 2 × 2 center hole.

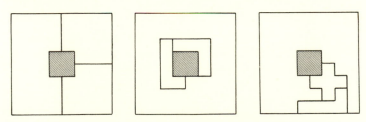

Figure 31. Impossible pentomino constructions for the 8 × 8 board.

of arranging the twelve pentominoes into a 6 × 10 rectangle. Excluding rotations and reflections, he found 2,339 basically distinct solutions! He also verified the results of Dana Scott's two programs.

Before leaving the subject of pentomino patterns, several special configuration problems seem worthy of mention:

1. The 64-square triangle, filled with the twelve pentominoes and the square tetromino. (Other tetrominoes may also be specified as the thirteenth piece.) One solution is shown in figure 32.

2. Another difficult configuration is the elongated cross of figure 33.

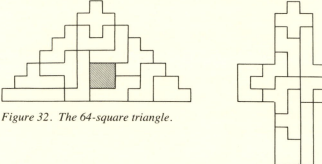

Figure 32. The 64-square triangle.

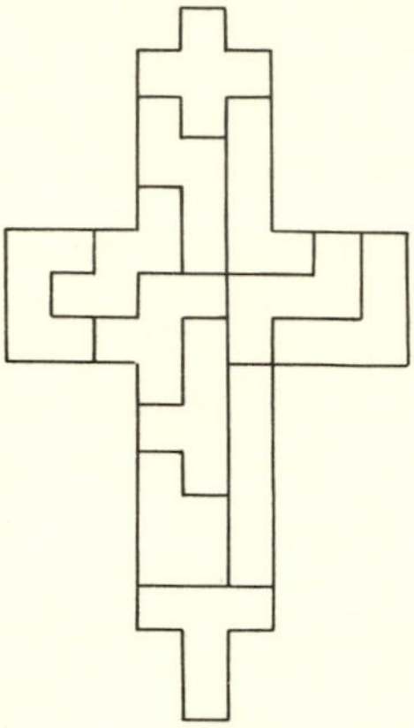

Figure 33. An elongated cross composed of pentominoes.

3. It was easily proved by Professor R. M. Robinson (who also first proposed the jagged square of chapter 6) that the 60-square pattern shown in figure 34 is incapable of holding the twelve pentominoes.

*Figure 34.
R. M. Robinson's
60-square region.*

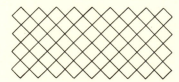

· In particular, there are 22 edge squares (including corners) in this pattern. If the pentominoes are examined separately, and the maximum number of edge squares each could contribute is listed, the total is only 21, as follows:

$$
\begin{array}{llll}
T = 1 & W = 3 & Z = 1 & L = 1 \\
U = 1 & X = 3 & F = 3 & P = 2 \quad \text{Total} = 21 \\
V = 1 & Y = 2 & I = 1 & N = 2
\end{array}
$$

This type of reasoning is used in solving jigsaw puzzles, where it is common practice to separate the edge pieces from the interior ones.

Two other interesting pentomino patterns are dealt with in chapter 4.

TETROMINO PROBLEMS

Unlike the pentominoes, the five distinct tetrominoes will not form a rectangle. To prove this, color rectangles twenty squares in area in checkerboard fashion, as indicated in figure 35. Four of the five tetrominoes will always cover an equal number of dark and light squares. However, the remaining tetromino always covers three squares of one color and one square of the other (see fig. 36). Hence, the five tetrominoes will cover a total of an *odd* number of dark squares and an *odd* number of light squares. However, the rectangles in figure 35 have ten squares of each color, and 10 is an even number.

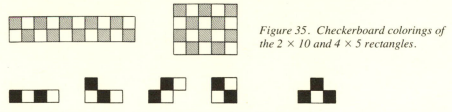

Figure 35. Checkerboard colorings of the 2 × 10 and 4 × 5 rectangles.

Figure 36. The 4 "balanced" tetrominoes and, at the far right, the "unbalanced" tetromino.

On the other hand, any of several different pentominoes can be combined with the five tetrominoes to form a 5 × 5 square. Two examples are given in figure 37. The reader is invited to investigate how many *different* pentominoes can be used in this manner.

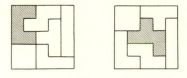

Figure 37. The tetrominoes may combine with a pentomino to form a 5 × 5 square.

MASONRY PROBLEMS

Robert I. Jewett, while a graduate student in mathematics at the University of Oregon, proposed the following problem: Is it possible to cover a rectangle with two or more dominoes

Chapter 2

Figure 38. A
rectangle of
dominoes, with
a fault line.

Figure 39.
Two fault-free
5 × 6 rectangles
of dominoes.

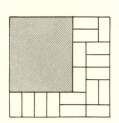

Figure 40. Extend-
ing a fault-free
5 × 6 rectangle to
a fault-free
8 × 8 rectangle.

Figure 41. A
6 × 6 rectangle
covered with
dominoes (not
fault free).

so that every *grid line* (that is, the lines, both horizontal and vertical, spaced at the width of one domino and extending perpendicularly between parallel edges) of the rectangle intersects at least one domino? For example, in the pattern of figure 38, the vertical grid line in the middle of the rectangle does not cut any dominoes. Thinking of dominoes as bricks, such a grid line represents a structural weakness. Jewett's problem is thus to find "masonry" patterns without "fault lines."

Many people who try this problem soon give up, convinced that there are no solutions. Actually, there are infinitely many, but the one using the smallest number of dominoes requires fifteen of them, arranged in a 5 × 6 rectangle. In fact, there are two basically different ways to form such a rectangle, as shown in figure 39.

It is not difficult to show that the minimum width for fault-free rectangles must exceed 4 squares. (The case of width 2 [a rectangle 2 squares wide] is easily ruled out as soon as covering one of the width-2 sides of the rectangle is attempted. Either one domino covers both squares forming the side, producing a fault line immediately, or else two separate dominoes cover the two squares, also producing a fault line. The reader is invited to rule out the cases of width 3 and width 4.) Hence, since 5 × 5 is an odd number of squares, while dominoes always cover an even number, the 5 × 6 rectangle is the smallest possible construction.

A 5 × 6 rectangle can be "extended" to the 8 × 8 checkerboard and still satisfy the fault-free condition. In particular, this can be done as indicated in figure 40. Surprisingly, there are no fault-free 6 × 6 squares. There is a remarkable proof for this: Imagine any 6 × 6 square covered entirely with dominoes. Such a figure contains eighteen dominoes (half the total number of squares) and ten grid lines (five horizontal and five vertical). One such covering (*not* fault free) is shown in figure 41. The grid lines that have not been intersected are indicated by heavy lines.

As noted earlier, such a figure is fault free only if each grid line intersects at least one domino. (Note that each domino is cut by *exactly* one grid line.) It will be shown that each grid line intersects an *even number* of dominoes. Hence, in the fault-free case, each grid line must intersect at least two dominoes. With ten grid lines, at least twenty dominoes would be intersected; but there are only eighteen of them on the 6 × 6 board.

It remains only to prove the assertion that each grid line cuts an even number of dominoes. Consider, for example, a vertical grid line. The area to the left of it is an *even* number of squares (6, 12, 18, and so on). The dominoes entirely to the left (that is, not intersected by the grid line) cover an even number of squares, since each domino covers two. The dominoes cut by the grid line must also occupy an even area to the left of it, because this area is the difference between two even numbers (the total area to the left and the area of the uncut dominoes to the left). Since each cut domino occupies one square to the left of the grid line, there must be an *even number* of dominoes cut by the grid line. Thus the proof is complete.

Similar reasoning shows that for a fault-free 6 × 8 rectangle to exist, every grid line must intersect *exactly* two dominoes. This is precisely what happens in the example shown in figure 42.

The most general result is the following: If a rectangle has an area that is an even number of units, and if both its length and width exceed 4, it is possible to find a fault-free domino covering of the rectangle, except in the 6 × 6 case. Actually, coverings for all larger rectangles can be extended from the 5 × 6 and the 6 × 8 rectangles, using a method of enlarging either the length or the width by 2. This procedure is easiest to explain by example. In figure 43 the extension of a 5 × 6 rectangle to a 5 × 8 rectangle is shown. Generally, to extend a rectangle horizontally by 2, a horizontal domino is placed next to each horizontal one at the old boundary, while vertical dominoes are shifted from the old boundary to the new, leaving an intervening space, which is filled with two horizontal dominoes for each vertical domino shifted.

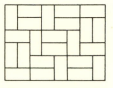

Figure 42. A fault-free 6 × 8 rectangle, where every grid line cuts exactly 2 dominoes.

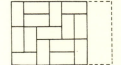 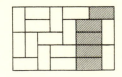

Figure 43. Extending the length (or width) of a fault-free rectangle by 2.

As a concluding problem in masonry patterns, the reader may find it interesting to study *trominoes* as bricks. In particular, what is the smallest rectangle that can be covered by two or more straight trominoes without any fault lines?

3 Where Pentominoes Will Not Fit

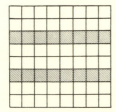

T, V, X, Z excluded

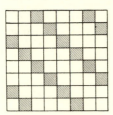

U, Y, I, L excluded

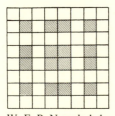

W, F, P, N excluded

Figure 44. Three patterns proving that 16 monominoes are sufficient to exclude any given pentomino from the checkerboard.

PENTOMINO EXCLUSION BY MONOMINOES

The fitting of pentominoes and other polyominoes into patterns of various shapes has been the theme of the problems thus far. This section's problems will have an opposite objective: namely, what must be done to keep a pentomino *off* the checkerboard? Specifically, for each of the twelve pentominoes, what is the least number of monominoes that can be placed on the 8 × 8 board so that a given pentomino can no longer be fitted onto the board? There are thus twelve distinct problems, one for each of the pentominoes. To solve these, it is necessary first to exhibit a way of placing a certain number of monominoes on the checkerboard so as to exclude the given pentomino and then to *prove* (by whatever combinatorial reasoning or tricks suggest themselves) that no fewer monominoes could have been used for the same purpose.

To begin with, it takes only three constructions to show that sixteen monominoes will always be *sufficient* to exclude any prescribed pentomino. (It will be seen later that, in six of the twelve cases, sixteen monominoes are also *necessary*.) These three placements are shown in figure 44. Under each board are the names of four pentominoes; none of these pieces can be fitted onto that particular partially occupied checkerboard. All twelve pentominoes are thus excludable by sixteen monominoes. In fact, three of the pentominoes (T, W, and F) are excluded by two out of the three partially covered boards above.

For some of the pentominoes (notably the U, W, Y, and L), many people experience difficulty in finding as few as sixteen locations for monominoes to exclude the given pentomino. However, once the patterns in figure 44 are found, the improvements of figure 45 are usually discovered quickly. Figure 45a exhibits a configuration of only twelve monominoes that succeeds in keeping the X pentomino off the board. Similarly, figure 45b uses only fourteen monominoes to keep off the I pentomino. However, much to the dismay of those who have pondered long and hard to find a way to do as well as this, neither of these results is the best possible. To keep the suspense to a minimum, the best configurations for the twelve pentominoes are shown in the nine constructions of figure 46.

Simple inspection will verify that the indicated numbers of monominoes, ranging from ten to sixteen, are indeed sufficient to exclude the specified pentominoes from the checkerboard. It remains to show that the number of monominoes

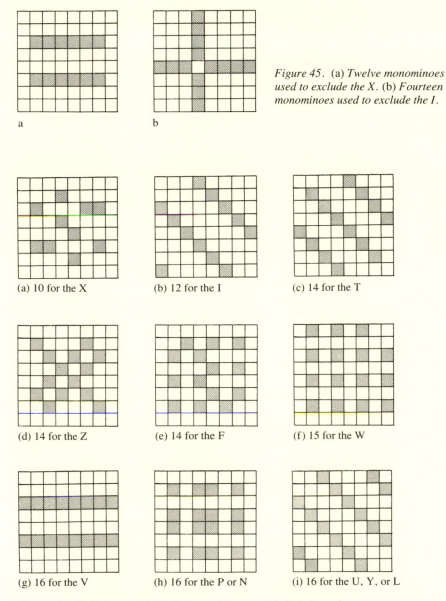

Figure 45. (a) *Twelve monominoes used to exclude the X.* (b) *Fourteen monominoes used to exclude the I.*

(a) 10 for the X   (b) 12 for the I   (c) 14 for the T

(d) 14 for the Z   (e) 14 for the F   (f) 15 for the W

(g) 16 for the V   (h) 16 for the P or N   (i) 16 for the U, Y, or L

Figure 46. Sufficient numbers of monominoes to exclude indicated pentominoes from the checkerboard.

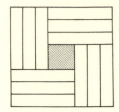

Figure 47. Construction used to show that at least 12 monominoes are necessary to exclude the I pentomino.

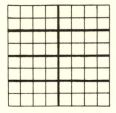

Figure 48. Construction used in the proof that at least 16 monominoes are needed to exclude the Y pentomino.

appearing in figure 46 are also *necessary*. All twelve necessity proofs are elaborations of the same basic combinatorial theme and will be discussed here in the order of increasing complexity. (The four most difficult proofs will be left somewhat incomplete.)

The simplest necessity proof is for the I pentomino. To show that twelve monominoes are necessary to keep the I pentomino off the 8×8 board consider the board decomposed as in figure 47, where there are twelve nonoverlapping 1×5 rectangles, each of which could independently hold an I pentomino. To exclude this piece there must be at least one monomino in each of the twelve 1×5 rectangles. Thus, a minimum of twelve monominoes is necessary to exclude the I pentomino. And, from figure 46b, twelve monominoes is also seen to be sufficient.

Actually, with the sole exception of the I pentomino, it is not possible to fit as many of a given pentomino onto the checkerboard as there are monominoes needed to keep that piece off the board. However, the proof just given is easily modified to treat the six pentominoes that actually require sixteen monominoes to keep them off the board.

Consider first the Y pentomino and decompose the checkerboard as in figure 48. To keep the Y pentomino off the board, it is certainly necessary to keep it out of each of the eight 2×4 rectangular regions. Moreover, a single monomino will *not* exclude the Y pentomino, there being only two inequivalent locations for the monomino: namely, a corner square or an interior square (fig. 49). Neither location will keep the rectangle free of Y pentominoes. Hence, at least two monominoes are needed in each of the eight rectangular regions, making a total of sixteen monominoes necessary. Figure 46i showed this number to be sufficient as well.

Figure 49. One monomino is insufficient to exclude the Y from a 2×4 rectangle.

The same decomposition of the board (fig. 48) works for the U, L, P, and N pentominoes also, since none of them is kept off a 2×4 rectangle by a single monomino (specifically, see fig. 50).

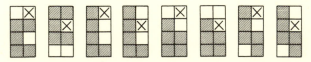

Figure 50. One monomino is insufficient to exclude the U, L, P, or N from a 2×4 rectangle.

Among the pentominoes requiring sixteen monominoes to exclude them, there remains only the V. For this case, decompose the checkerboard into quadrants (fig. 51). To show that each quadrant must contain at least four monominoes, suppose there were a quadrant with only three monominoes. Then in that section, at least one *rank* (horizontal row) and at least one *file* (vertical row) must be vacant. A rank of four squares intersecting a file of four in length will always be able to hold a V pentomino, as shown in figure 51, where the three inequivalent cases (outside rank with outside file, inside rank with inside file, and inside rank with outside file) are illustrated. Hence, each of the four quadrants must contain at least four monominoes in order to exclude the V, making a total of a minimum of sixteen on the entire board.

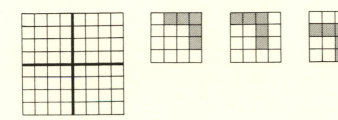

Figure 51. At least 16 monominoes are needed to exclude the V.

The next most simple case is that of the T pentomino. This time, decompose the 8 × 8 board as in figure 52a. There are five areas, an 8-square region in the center, and four congruent 14-square regions around the edge. The central region cannot exclude the T pentomino with but a single monomino, as a glance at the two inequivalent locations for the monomino (figs. 52b and 52c) indicates. Hence, the central region must contain at least two monominoes. To show that each outside region must contain at least three monominoes, for a total of at least 4 × 3 + 2 = 14 monominoes on the entire board, suppose the outside area could exclude the T pentomino with only two monominoes. Decomposing it as in figure 52d, the region covered by the T pentomino would have to contain a monomino, and there would be only one monomino left for the other subregion. The square indicated by a cross is the

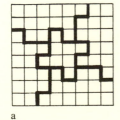

a

b

c

d

Figure 52. Constructions used in the proof that at least 14 monominoes are needed to exclude the T.

only monomino location that keeps T pentominoes out of the 9-square subregion. However, as figure 53 shows, each of the five assignments of a monomino to the T portion leaves a loophole somewhere in the 14-square region. Thus, three monominoes in each of these regions are necessary, and the fourteen monominoes that are shown to be sufficient in figure 46c are also seen to be necessary.

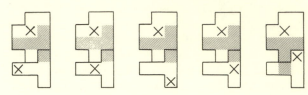

Figure 53. Further constructions in the T-exclusion proof.

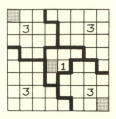

Figure 54. Decomposition used in the W-pentomino exclusion proof.

The remaining four pentominoes are more difficult to treat. Of these, the W will be considered first and the 8 × 8 board decomposed as shown in figure 54. It is rather easy to demonstrate that each of the five major subregions requires at least the number of monominoes listed on it to exclude the W pentomino. However, this merely leads to the conclusion that thirteen monominoes are necessary, while the objective is to show that, in fact, fifteen are necessary.

Suppose thirteen monominoes were sufficient. Then, no matter how figure 54 is rotated, the indicated number of monominoes must exclude the W. Superposition of all the constraints of rotation and reflection leads to figure 55. Each of the four "corner" regions would require three monominoes, and the dark region in the center, being congruent to one of the corner regions of figure 54, requires three more, for a total of fifteen. Since this argument began with the assumption that thirteen would be adequate, a contradiction has been reached, and thirteen has been shown to be insufficient. A lengthy refinement of this argument would show that fifteen are necessary. From figure 46f, this number is also known to be sufficient.

Figure 55. Superposition of constraints imposed by rotations and reflections of Figure 54.

The remaining 3 pentominoes (X, F, and Z) are analogous to the W in their treatment. First, a single illustration can be used to arrive within two of the actual number of sufficient monominoes. For example, see figure 56.

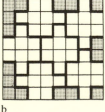
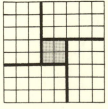

Figure 56. Decomposition showing monominoes needed to exclude (a) the X, (b) the F, (c) the Z.

a   b   c

It is seen in 56a that at least eight monominoes are needed to exclude the X; in 56b, that at least twelve are needed to exclude the F; and in 56c, that at least twelve are needed to exclude the Z.

To show that at least nine monominoes are needed against the X, consider figure 57a. If eight monominoes were sufficient, they would have to lie in the 8 X regions of 57a no matter how that pattern were rotated and reflected on the checkerboard. That is, all eight would have to be in the *unshaded* portion of figure 57b. However, the large central region of 57b actually requires at least two more monominoes to keep out the X, contradicting the sufficiency of the eight monominoes. Again, further refinement of this argument will show that ten are necessary.

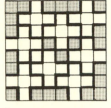

a

For the F, consider figure 58a. The regions require the indicated number of monominoes to exclude the F. To show this for the area with the number 3, consider its decomposition in figure 58b. If two monominoes were sufficient, one would have to be in the F subregion. The only location for a monomino that would keep an F out of the other subregion is at the square indicated by an *x*. However, any square in the F subregion combined with the *x* square fails to exclude the further occurrence of F's.

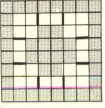

b

The region in figure 58a with the two check marks could hold an F with either checked square excluded. When all the rotations and reflections of 58a are considered, the conclusion is reached that if twelve monominoes would suffice to exclude F's, they would all have to be in the unshaded portion of figure 58c. However, at least two more monominoes would then be needed in the central, shaded region, proving that twelve monominoes are insufficient. Further refinement of this argument leads to the conclusion that fourteen monominoes are both necessary and sufficient.

Figure 57. Constructions used to prove that 8 monominoes are insufficient to exclude the X.

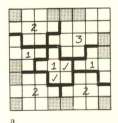

a

b

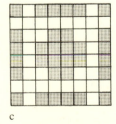

c

Figure 58. Constructions used to prove that 12 monominoes are insufficient to exclude the F.

There remains only the case of the Z pentomino. In figure 59a each of the four "corners" is readily shown to require at least three monominoes; this is accomplished by the usual expedient: a Z is drawn in the region and it is observed that one monomino within the Z and another outside it is insufficient to exclude Z pentominoes.

By rotating and reflecting figure 59a in all possible ways, it is seen that if twelve monominoes were sufficient to exclude the Z pentomino, then all twelve would lie in the unshaded portion of figure 59b. However, the shaded portion of 59b actually can contain four more Z pentominoes simultaneously. Hence, at least thirteen monominoes are needed.

The argument that makes it possible to arrive at fourteen as the necessary and sufficient number of monominoes to exclude the Z pentomino will now be sketched. First, divide the checkerboard into the four congruent quarters shown in figure 59c. Suppose that thirteen monominoes were sufficient to exclude the Z pentomino (that number already known to be necessary). Then, in the hypothetical 13-monomino configuration, at least one of the four quarters will contain a minimum of four monominoes (since four whole numbers, all less than 4, cannot add up to 13).

Figure 59.
Constructions used to prove that 12 monominoes are insufficient to exclude the Z.

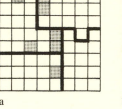
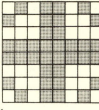
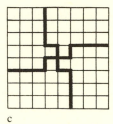

a   b   c

Rotate the hypothetical 13-monomino configuration so that the upper left-hand quarter has at least four monominoes, leaving nine at most for the rest, and divide the remaining area into the two regions shown in figure 60. Since the smaller, shaded region must contain at least three monominoes in order to exclude the Z, six at most are available for the large, unshaded region. Proving that six monominoes are insufficient to exclude the Z from this large region will thus complete the proof. This will be accomplished by showing that the assumption that six monominoes suffice leads to a contradiction.

Figure 60.
Construction used to contradict the assumption that 13 monominoes suffice to exclude the Z.

If six monominoes were enough, they would lie within the two unshaded regions of figure 61a, since each of those regions requires three monominoes. They would also lie within the two unshaded regions of figure 61b, as well as the two unshaded regions of figure 61c, d, and e. Combining these results, six monominoes, three in each of the two unshaded regions shown in figure 61f, must keep the Z pentomino off the entire 34-square region.

Consider the upper portion. Three monominoes in the un-

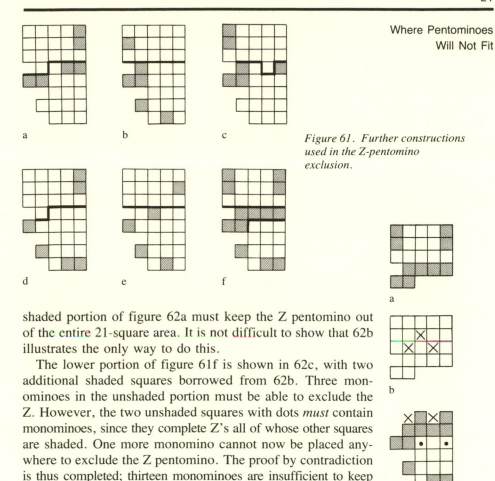

a   b   c

Figure 61. Further constructions used in the Z-pentomino exclusion.

d   e   f

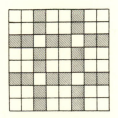

a

b

c

Figure 62. Final constructions in the Z exclusion.

shaded portion of figure 62a must keep the Z pentomino out of the entire 21-square area. It is not difficult to show that 62b illustrates the only way to do this.

The lower portion of figure 61f is shown in 62c, with two additional shaded squares borrowed from 62b. Three monominoes in the unshaded portion must be able to exclude the Z. However, the two unshaded squares with dots *must* contain monominoes, since they complete Z's all of whose other squares are shaded. One more monomino cannot now be placed anywhere to exclude the Z pentomino. The proof by contradiction is thus completed; thirteen monominoes are insufficient to keep the Z pentomino off the checkerboard; fourteen monominoes are therefore necessary and, as seen earlier (fig. 46d), also sufficient.

MODIFICATIONS AND GENERALIZATIONS

A number of interesting variants of the problems in the preceding section can be considered. For example, one can try to find the minimum number of dominoes to be placed on the 8 × 8 checkerboard so as to exclude all twelve pentominoes *simultaneously*. The solution to this problem is shown in figure 63. If each of these twelve dominoes is regarded as two monominoes, it is evident that twenty-four monominoes suffice to exclude all twelve pentominoes; and no solution with fewer than twenty-four monominoes is believed to exist.

If one asks for the minimum configuration of monominoes excluding at least eleven of the twelve pentomino shapes, the

Figure 63. Twelve dominoes must be used to exclude all 12 pentominoes simultaneously.

Chapter 3

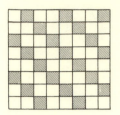

*Figure 64.
Twenty-one
monominoes exclude
all but the
W pentomino.*

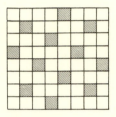

*Figure 65.
Twelve is the
minimum number
of monominoes
needed to exclude
2 pentominoes
(the X and the I).*

answer is believed to be as shown in figure 64. Here, twenty-one monominoes suffice to exclude all pentominoes except the W.

Another multiple-pentomino exclusion problem is to find the configuration using the fewest monominoes possible to exclude at least two of the pentominoes. Since a glance at figure 46 will reveal that fewer than twelve monominoes can hope to exclude only the X pentomino, it is remarkable that twelve monominoes *can* be placed, as in figure 65, to exclude both the X and I pentominoes, thereby solving the problem. Altogether, there are 4,095 problems involving the use of monominoes to exclude individual pentominoes, pairs of pentominoes, triples, and on up to exclusion of all twelve pentominoes at once; this leaves many additional exercises for the ambitious reader.

It is also instructive to see how many monominoes are needed to exclude the various polyominoes of lower order than the pentominoes from the checkerboard. The minimal configurations are shown in figure 66, but the proofs that they are indeed minimal will be left to the reader.

Notice how these patterns already foreshadow most of those in figure 46. Naturally, a monomino configuration excluding a given polyomino will certainly exclude any extension of that polyomino, and every pentomino is the extension of at least one, and often several, tetrominoes.

A further insight into the problem of excluding pentominoes from the 8 × 8 board is obtained by studying the related problem of using monominoes to exclude the various pentominoes from an *infinite* planar array of squares. Four configurations illustrate the minimal solutions for all the pentominoes. These are shown in figure 67.

In figure 67a, it is shown that only one-fifth of the squares need be covered with monominoes to keep off the I and X pentominoes. In figure 67b, one-fourth of the squares must be covered to exclude the T, U, Y, and L pentominoes; as 67c shows, one-fourth of the squares must also be covered to exclude the W, F, P, and N pentominoes. Finally, in 67d, a full one-third of the squares are covered to exclude the V and Z pentominoes.

The methods used in this chapter involved the strenuous application of logic and ingenuity to the problems of excluding specified pentominoes with a minimum number of monominoes. However, there are other problems that resist solution by ingenuity but that can be solved, at least in principle, by examining all possible cases. Such a technique is described in the next chapter.

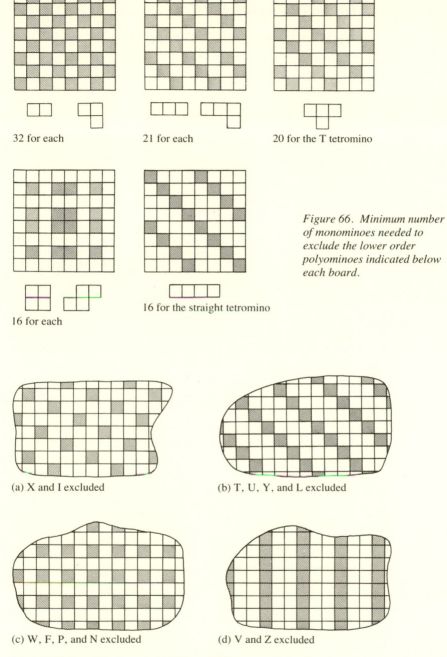

32 for each   21 for each   20 for the T tetromino

Figure 66. Minimum number of monominoes needed to exclude the lower order polyominoes indicated below each board.

16 for each

16 for the straight tetromino

(a) X and I excluded

(b) T, U, Y, and L excluded

(c) W, F, P, and N excluded

(d) V and Z excluded

Figure 67. Repetitive patterns which exclude pentominoes from the infinite plane.

4 Backtracking and Impossible Constructions

BACKTRACK PROGRAMMING

The method of proof to demonstrate impossible constructions is an exhaustive search procedure called *backtrack*, the idea of which is to pursue each possible line of progress in turn until it leads either to success or to a blind alley. In the latter case, one then "backtracks" to the next possible procedure. After a preliminary discussion this technique will be applied to two very difficult pentomino problems.

Backtracking is illustrated conceptually by the maze-threading problem (fig. 68); the man in the maze (starting at X) adopts the escape strategy of running his right hand along the wall at all times; he follows the path indicated by the dots in figure 68 until he is finally outside. The success of his strategy depends upon the construction of the maze walls. This incorporates the essential backtrack feature that the same ground need not be searched repeatedly (a maze with the wrong wall connectivity will not yield to this method of solution); it also necessitates that the man recognize when he has finally succeeded. If he does not realize he has escaped, he will continue to follow his right hand along the doorjamb all the way around the *outside* wall of the building and back in the other side of the maze entrance.

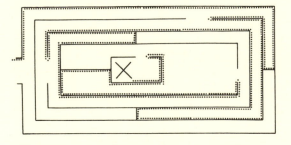

*Figure 68.
Escaping from a
maze.*

Another example of backtrack programming is the classic combinatorial problem of placing eight queens on the checkerboard so that no two can mutually "attack"; that is, no two can be in the same rank, file, or diagonal. Rather than examine *all* ways of placing eight queens on the board (the number is a staggering 4,426,165,368, computed by a method explained in chapter 5), it suffices to restrict each queen to a single row of the board and to place them, one at a time, in unattacked squares. The first sequence of moves is shown in figure 69a. To make the process converge fairly rapidly, the first queen is placed in a "typical" (middle) square of the first (bottom) row. The second queen is in the *first* (farthest right) available square of the second row, and the third queen is on the first available (unattacked) square of the third row. Similarly, the fourth queen is on the first (farthest right) available square of the fourth row, and similarly with the fifth queen on the fifth row and the sixth queen on the sixth row.

Now, however, there is no available square for a queen *anywhere* on the seventh row, so that it is necessary to "backtrack" to the position in figure 69b. This can be "extended" to the situation in figure 69c. Now, however, there is no available square in the seventh row, and it is necessary to backtrack all the way to figure 69d.

This can be extended to the situation in figure 69e, but from this near miss one must retreat to figure 69f, which can be extended to figure 69g. From here it is necessary to backtrack to figure 69h; advance can then be made to figure 69i. Now

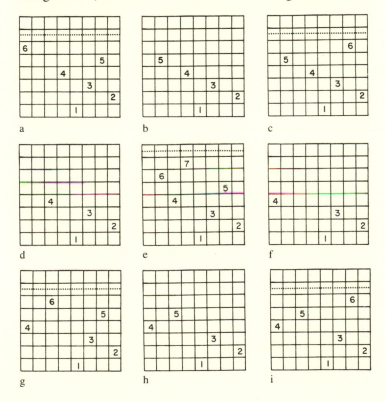

Figure 69.
Backtrack program-
ming for the
8-queen problem.

Chapter 4

it is necessary to go all the way back to the queen in the third row (fig. 69j). It has taken this long to establish that no solution exists with the first three queens as originally placed. Now, however, it is possible to advance, one queen at a time, all the way to the solution in figure 69m. Although the path was somewhat laborious, it has led to success in an infinitesimal fraction of the time required to examine some four billion

Figure 69.
continued

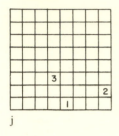

j

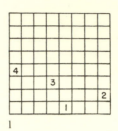

k   l

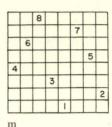

m

Figure 70. Most symmetrical solution to the 8-queen problem.

cases. Moreover, the backtracking procedure can be continued to determine all possible solutions in a reasonable length of time. The ambitious reader who wishes to attempt this should remember that four distinct locations for the *first* queen are involved if all solutions are to be found. The solution possessing the most symmetry is shown in figure 70. There are ninety-two answers altogether, which reduce to twelve distinct cases if all rotations and reflections of the checkerboard are taken into account. In attempting to find all the solutions by backtrack, there is a considerable saving in time if these symmetries are accounted for in advance. Thus, after all answers have been found with the first queen located as in figure 69m, no further partial solutions need be considered that have a queen on any edge four squares from any corner.

Although backtrack is not "elegant" in the way of a clever mathematical proof, it is an invaluable tool in the study of combinatorial problems. Moreover, the systematic pattern of advance and retreat until a solution is found, or the possibilities are exhausted so that the nonexistence of a solution is established, makes backtrack very well suited for the programming of an electronic digital computer. The digital computer is programmed by transforming such a pattern into numerical form. The computer then, according to instructions, manipulates the numerical progression until a solution is found. Unlike a human problem solver, the modern digital computer does routine, repetitive calculations with incredible rapidity but fails to notice any shortcuts or patterns that had not already occurred to the programmer who gave the machine its detailed instructions. The challenge, then, is to formulate the question efficiently as a backtrack problem, after which the computer takes over the more tedious task of examining all the cases.

THE HERBERT TAYLOR CONFIGURATION

Impossibility proofs for figure 71, the configuration proposed by the mathematician Herbert Taylor, now at the Center for Communications Research in La Jolla, California, were discovered independently by John G. Fletcher of the University of California at Berkeley, and Spencer Earnshaw, at the time a student at Santa Monica City College in California. The proof presented here is a simplification of these two previous ones.

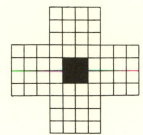

Figure 71. The Herbert Taylor Configuration.

Suppose it would be possible to cover the configuration of figure 71 with the twelve distinct pentominoes. Then, at no stage along the way could a region of the "board" be isolated unless it contained a multiple of five squares. This will lead to a dead end.

First, consider the location of the X pentomino in the hypothetical covering. There are only two possibilities (ignoring positions symmetric to these as being equivalent), as shown in figure 72.

In Case I, consider all possible locations for the I pentomino. No matter where it is placed, a region is isolated with a number of squares not a multiple of 5. Hence, Case I is a blind alley.

In Case II, there are two inequivalent locations for the I pentomino that leave the remaining regions with numbers of squares divisible by 5, as shown in figure 73.

In Case IIA, consider the possible ways the pentomino covering the "dotted" square might enter the "east wing" of the figure. There are three distinguishable situations (fig. 74).

Chapter 4

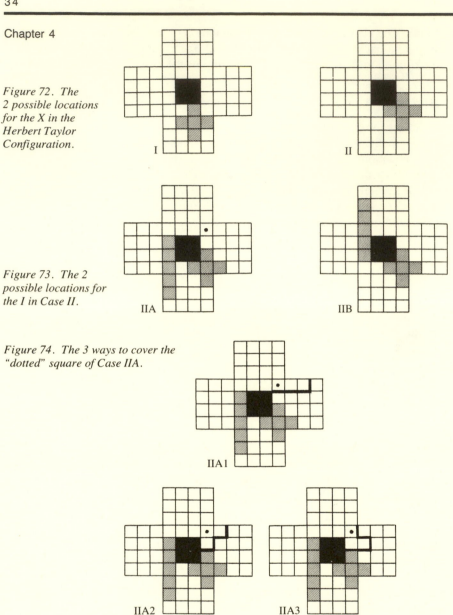

*Figure 72. The
2 possible locations
for the X in the
Herbert Taylor
Configuration.*

I

II

*Figure 73. The 2
possible locations for
the I in Case II.*

IIA

IIB

*Figure 74. The 3 ways to cover the
"dotted" square of Case IIA.*

IIA1

IIA2

IIA3

In Case IIA1, the south and east wings are congruent, and
either can be filled by the P and L pentominoes or by the W
and Y pentominoes, but in no other way. Thus P, L, W, and
Y must be used (for example, as in fig. 75), and the pentomino covering the dot is either the N or the V. However, a
search for a place to put the T pentomino leads only to the
location shown in figure 75, which also necessitates that the

V cover the dot. Now, however, the P pentomino is needed next to the T, but it has already been used. Thus Case IIA1 is eliminated. (The numbers on the pentominoes in figure 75 refer to the order in which they were placed on the "board.")

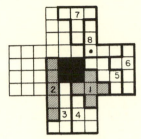

Figure 75. Exploring Case IIA1.

In Case IIA2, the south wing, as before, can contain either the pair P and L or the pair W and Y; while the east wing can contain either W and P or L and P. The only consistent choice for *both* regions is W, Y in the south and L, P in the east (see fig. 76). The pentomino covering the dot must be the F, in either of two orientations. Next, the N pentomino can be consistently fitted only as shown (fig. 76), and there remains no consistent location for the T pentomino, thus eliminating Case IIA2.

Figure 76. Exploring Case IIA2.

In Case IIA3, the south wing again may contain either the pair P, L or the pair W, Y, while the east wing may contain either V, F or U, L. Next, there are four possibilities for the T pentomino (avoiding obvious dead ends), as shown in figure 77. In IIA3a, the dotted square and the neighboring asterisked ones cannot be extended to form a pentomino, because W already has been used, and N would isolate a block of two squares. In IIA3b, the N must be used to cover the dot, and there remains no consistent location for the Z. For IIA3c, the L must cover the dot, and then the square to the left of the dot cannot be covered. In IIA3d, whether V, F, or U, L are chosen to cover the east wing, only Z can be used to cover the dot, and the construction clearly cannot be completed. Thus IIA3 is impossible, which completes the elimination of Case IIA.

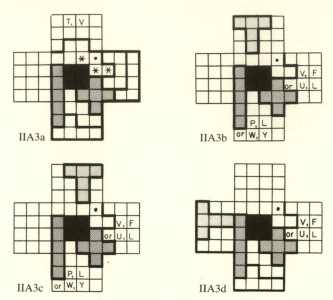

Figure 77. The 4 possible locations for the T in Case IIA3.

Case IIB remains. Here, there are exactly four possible locations for the N pentomino, as shown in figure 78. In IIB1, the only consistent location for the T pentomino is as shown

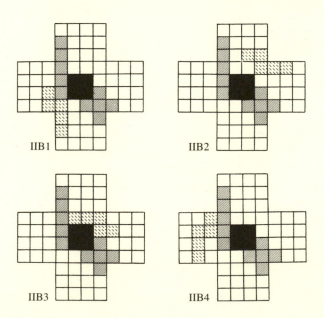

Figure 78. The 4 locations for the N in Case IIB.

in figure 79. The north wing can then be completed only by
use of the P pentomino. Without the P, the south wing must
be filled in with W, Y and the west wing with V, Z. The
asterisked square of figure 79 cannot then be covered by any
remaining pentomino, which settles case IIB1.

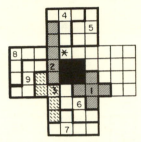

*Figure 79. The only position for the
T in Case IIB1.*

For IIB2, the north wing must be filled with U, L, and (since
L is no longer available) the east wing must contain W, Y.
The only consistent location for the T pentomino is then in
the west wing (see fig. 80), after which the Z pentomino will
not fit anywhere. Thus Case IIB2 is also eliminated.

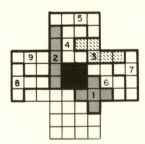

*Figure 80. The only consistent location
for the T in Case IIB2.*

In IIB3, the east wing may be occupied either by U, L or
F, V; and the north wing by U, Y or P, T (see fig. 81). (If
P is used in the north wing with V, Y, Z, or L, then there is
no suitable location left for the T.)

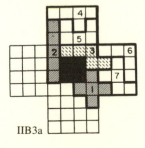

IIB3a

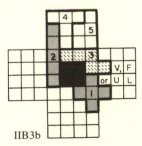

IIB3b

*Figure 81. The U
and Y or the P and T
must be used to fill
the north wing in
Case IIB3.*

In Case IIB3a, the only consistent location for the T pentomino is in the west wing (see fig. 82), which then leaves no legitimate position for the Z. Considering IIB3b, there are two possible locations for the Z pentomino, as shown in figure 83. In IIB3bi, there is no region left where the W pentomino will fit. In Case IIB3bii, the location of the Z requires the V in the west wing, leaving U, L to fill the east wing. There is only one location left for the Y pentomino, as shown, after which the F pentomino does not fit at all. This eliminates Case IIB3b, which completes Case IIB3.

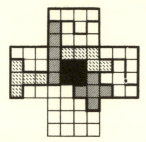

Figure 82.
Elimination of Case
IIB3a.

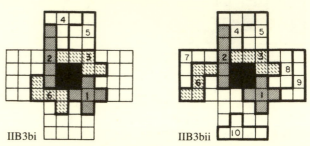

IIB3bi   IIB3bii

Figure 83. The two possibilities for locating
the Z in Case IIB3b.

Finally, Case IIB4 is eliminated by observing that the T must be placed in the north wing, above the I, which forces the P next to the T and the Y underneath; the east wing cannot now be filled. This completes Case IIB, completing Case II, and answering the original question of the existence of a pentomino covering for figure 71 in the negative. A schematic diagram of the cases and subcases just considered is given in figure 84.

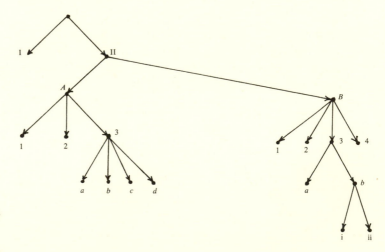

Figure 84.
Branches used in the
impossibility proof.

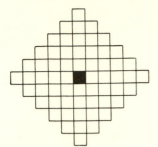

Figure 85. R. M. Robinson's jagged square.

THE JAGGED-SQUARE CONFIGURATION

A backtrack "solution" to the "jagged square" proposed by R. M. Robinson (fig. 85) was also discovered by Spencer Earnshaw. It is even longer and more intricate than the preceding solution and will only be sketched briefly here, in eight steps.

1. Only eight of the twelve pentominoes (W, X, Y, F, I, L, P, N) can possibly occupy "corners" in figure 85.

2. There are seventy ways $\left(\dfrac{8!}{4!4!} = 70\right)$ to select four of these pentominoes to fill the four corners. (The formula for the number of ways to select eight things, four at a time, comes from Theorem 4 in chapter 5. The meaning of the symbol "$n!$", called *n factorial*, is explained in the statement of Theorem 2 in chapter 5). The following table gives the maximum number of edge squares (*not* including corner squares) each pentomino can occupy when it is and when it is not a corner piece. For each set of corner pieces, the corresponding maximum number of coverable edges is obtained readily from table 1.

3. In figure 86, the four kinds of squares of figure 85 are depicted clearly. Topologically, the corner squares, edge

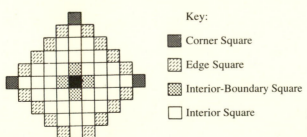

Key:

■ Corner Square

▨ Edge Square

▩ Interior-Boundary Square

□ Interior Square

Figure 86. Identification of corner, edge, interior-boundary, and interior squares.

TABLE 1

Pentomino	Corner Squares	Max. No. of Edge Squares
W	1	2
	0	3
X	1	2
	0	2
Y	1	1
	0	2
F	1	2
	0	2
I	1	0
	0	1
L	1	0
	0	1
P	1	1
	0	2
N	1	1
	0	2
T	0	1
U	0	1
V	0	1
Z	0	1

squares, and interior-boundary squares, all taken together, form the "total boundary" of the configuration, and the basic idea of Earnshaw's solution is to relate coverings to this "total boundary" of twenty-four squares.

Specifically, there are only four pentominoes (V, Z, I, and L) that can occupy an interior-boundary square and also cover their maximum number of edge or corner squares simultaneously.

4. The V and I pentominoes cannot simultaneously occupy both an edge or corner square and an interior-boundary square without violating the *multiple-of-5* rule for the remaining regions; this rule requires that the number of squares in an isolated region must be a multiple of 5 in order for that region to be exactly coverable by pentominoes. This is illustrated in figure 87; the I pentomino

has been placed in its two representative locations, and the various inequivalent positions of the V pentomino occupying both an edge and an interior-boundary point are shown. In all sixteen cases, the multiple-of-5 rule is violated.

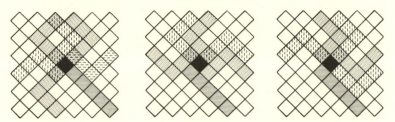

Figure 87. The 16 ways for the V and I simultaneously to occupy an edge or corner square and an interior-boundary square all violate the "multiple-of-5" rule.

5. There are fifteen remaining ways of selecting two from the V, Z, I, and L pentominoes to cover both an edge or corner and an interior-boundary square to be studied. (There are six ways to select two of these four pieces, and, in general, several ways of placing the selected pair on the "board.") A preliminary count of boundary squares shows that one of these situations would have to occur. However, it is found that only the pair I, Z has any hope of success.

6. The cases where I and Z each cover both an edge or corner square and an interior-boundary square are studied. Numerous side results (for example, that F and X are two of the four corner pieces) are obtained from this study. Finally, it is shown that I and Z must cover exactly one interior-boundary square and one edge or corner square, while another piece (either T, V, or L) covers the remaining two interior-boundary squares.

7. The assumption that the V covers the remaining two interior-boundary squares (see fig. 88) leads in all cases to a contradiction.

8. If V is *not* used as the piece covering the remaining two interior-boundary points, contradictions are also reached. This step concludes the impossibility proof associated with figure 85.

The reader is invited to attempt to put the flesh on the very meager skeleton of the proof that has just been given. Even more significantly, he or she is encouraged to discover a shorter procedure for arriving at the same conclusion.

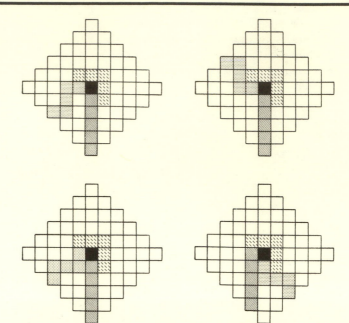

Figure 88. The V cannot cover the remaining 2 interior-boundary squares.

Even if the monomino is moved to another interior location, no solution to placing the twelve pentominoes on the remaining sixty squares has been found. The closest approximation known to the original pattern is the one given in figure 89.

A near solution to the jagged-square construction has been discovered by the recreational-mathematics fan J. A. Lindon of Surrey, England. In it, the monomino has been moved from the center to the edge (fig. 90).

Figure 89. A solution to a close approximation of the jagged-square problem.

Figure 90. The jagged square can be filled with pentominoes if the hole is moved to the edge.

Some Theorems about Counting

IT HAS OFTEN been said that the most basic of all mathematical operations is *counting*, and the late great German mathematician Hermann Weyl once wrote that underlying almost all mathematical identities is the principle that if an enumeration of the same objects is performed in two different ways, the results must be equal.

Throughout earlier chapters, questions have arisen as to the number of ways certain objects can be fitted into specified "containers," or patterns. There are also frequent questions in this book, and in mathematics generally, involving how many objects with certain specified characteristics exist, and involving problems of determining the number of configurations and patterns of a particular type. All of these are *finite enumeration* problems, and, like combinatorial geometry, with which this book is primarily concerned, they belong to a branch of mathematics called *combinatorial analysis*, which is the study of combinations of numbers or quantities of things. The ever-increasing importance of digital computers in modern technology has revived a widespread interest in combinatorial analysis, a subject that also has had important applications in such modern scientific fields as circuit design, coded communications, traffic control, crystallography, and probability theory.

The present chapter is intended to serve as an introduction to combinatorial analysis and, particularly, that part of the field concerned with *enumeration* (that is, counting). This material, while it is more difficult and requires greater concentration than that of earlier sections, will serve not only to enhance the readers' understanding and appreciation of the remainder of the book, but also to give them some facility in the formulation and solution of such problems wherever they may be encountered. However, for those who wish to concentrate on the recreational aspect of polyominoes, this chapter can be skipped and the rest of the book enjoyed without it.

COMBINATIONS

There are ten ways to pick a number from 0 through 9. There are $10^2 = 100$ ways to pick a *pair* of numbers, each digit one of the numerals from 0 through 9, since these pairs may be regarded as the numbers 00, 01, 02, 03, ..., 97, 98, 99. Similarly, there are $10^3 = 1,000$ ways to pick a *triple* of numbers, each digit again one of the numerals from 0 through 9, since the triples may be regarded as all the numbers from 000 to 999. In general, there are 10^n ways to pick an *n*-tuple of numbers, each digit a number from 0 to 9. In fact, there are 10^n ways to pick an *n*-tuple of *symbols* from some set of ten distinct symbols, since it is only a matter of nomenclature as to whether the symbols are called 0, 1, 2, ..., 9; *A, B, C,* ..., *J*; $a_1, a_2, a_3, ..., a_{10}$; or anything else.

If the basic *set*, or group of symbols from which the *n*-tuple is to be picked, consists of *k* symbols, and an ordered sequence of length *n* is formed—where each symbol can be any one of the *k* basic symbols—there are k^n such *n*-tuples. If the basic symbols are 0 and 1, and all possible sequences of length 4 are to be formed, $n = 4$ and $k = 2$, so that there are $k^n = 2^4 = 16$ such sequences. These combinations can be used to represent the decimal-system numbers from 0 to 15 (to the left of the equal signs in the table below) in the binary number system, which is based on the two digits 0 and 1 and used in computer mathematics, as follows:

$$
\begin{array}{llll}
0 = 0000 & 4 = 0100 & 8 = 1000 & 12 = 1100 \\
1 = 0001 & 5 = 0101 & 9 = 1001 & 13 = 1101 \\
2 = 0010 & 6 = 0110 & 10 = 1010 & 14 = 1110 \\
3 = 0011 & 7 = 0111 & 11 = 1011 & 15 = 1111
\end{array}
$$

A still more general formulation about combinations is the following theorem.

Theorem 1. Suppose there are *n* sets, where the first set contains k_1 objects, the second set contains k_2 objects, and so on, with the *n*-th set containing k_n objects. The number of different sequences of *n* objects, with the first object chosen from the first set, the second object from the second set, and so on, with the *n*-th object from the *n*-th set, is the product $k_1 k_2 ... k_n$. (In particular, if all *n* sets of objects are the same, then $k_1 = k_2 ... = k_n = k$, and the number of combinations is k^n.)

Proof. An object may be picked from the first set in any of k_1 ways. For each choice an object may be picked from the second set in k_2 ways, giving $k_1 k_2$ ways to make a selection from the first two sets. For each of these $k_1 k_2$ selections, an object may be picked from the third set in any of k_3 ways, giving $k_1k_2k_3$ ways to make the selection from the first three sets. Proceeding in this fashion, there are $k_1k_2k_3 \ldots k_n$ ways to make a selection from each of the n sets.

As an illustration of Theorem 1, the automobile registration plates in a hypothetical state consist of two letters (from the standard 26-letter alphabet), followed by four numerals (from the usual set of ten). Then the number of possible license plates in that state is $26 \cdot 26 \cdot 10 \cdot 10 \cdot 10 \cdot 10 = 6,760,000$ (the dots between the numerals are equivalent to multiplication signs).

A more difficult question is the number of *unordered* combinations of n symbols from a basic "alphabet" of k symbols. By an unordered combination, it is meant that the order in which the n symbols are arranged is to be disregarded. For example, if four symbols are taken from the binary number system, which consists of only the two digits 0 and 1, but no attention is paid to order, there are only five distinct cases: namely, 0000, 0001, 0011, 0111, and 1111. This contrasts with the sixteen distinct *ordered* cases previously listed. When two identical dice are thrown, the number of distinguishable configurations is not thirty-six, but only twenty-one, since, for example, (1, 4) and (4, 1) are not distinguishable. (Of course, there are only eleven different possible dice *totals*, these being the eleven numbers from 2 to 12 inclusive. However, the total of 7, for example, is achieved in three distinguishable ways, as $3 + 4$, $2 + 5$, and $1 + 6$.) The solution to the problem of counting the number of unordered combinations will be reserved until the next section, since additional concepts must first be introduced.

Exercises

1. In a typical 7-digit telephone number, the first two digits are chosen from 2 through 9 inclusive, while the remaining five digits can be any number from 1 to 0 inclusive. How many possible telephone numbers are there? (NOTE: *The answers to all Exercises in this chapter will be found in Appendix A.*)

2. There are six classes in a country school, with 12 students in the first grade, 15 in the second, 9 in the third, 13 in the fourth, 11 in the fifth, and 12 in the sixth grades. One student from each class is to be appointed to the safety committee. In how many ways can this committee be constituted?

Permutations, Factorials, and Binomial Coefficients

A set of k distinct objects is given, and a first object, a second object, and finally a k-th object are to be picked from it. In how many ways can this be done? For example, in how many different orders can a deck of fifty-two cards be dealt out; and in how many distinct ways can the letters A, B, C, D, E be arranged? The answer is given by the following theorem.

Theorem 2. The number of different ways in which k objects can be assigned an *ordering* (that is, first, second, third, through k-th) is the product $1 \cdot 2 \cdot 3 \ldots (k-1) \cdot k$. (This product is denoted by $k!$, and, as mentioned earlier, is called k *factorial*.)

Proof. The choice of the first object can be made in k ways. This leaves $k-1$ possibilities for the second object, $k-2$ choices for the third, and so on. Hence, the complete ordering can be done in exactly $k(k-1)(k-2) \ldots 2 \cdot 1 = k!$ distinct ways.

In table 2 are the values of $k!$ for $k = 1$ through 20.

TABLE 2

k	$k!$	k	$k!$
1	1	11	39,916,800
2	2	12	479,001,600
3	6	13	6,227,020,800
4	24	14	87,178,291,200
5	120	15	1,307,674,368,000
6	720	16	20,922,789,888,000
7	5,040	17	355,687,428,096,000
8	40,320	18	6,402,373,705,728,000
9	362,880	19	121,645,100,408,832,000
10	3,628,800	20	2,432,902,008,176,640,000

Since $3! = 6$, there are six ways to arrange the letters A, E, T; namely, AET, ATE, EAT, ETA, TAE, TEA. Similarly, there are $5! = 120$ ways to arrange the letters A, B, C, D, and E to form "words." The reader may find it amusing to look for sets of k letters whose permutations include as many English words as possible, for $k = 2, 3, 4, 5,$ and 6. For $k = 4$, it is difficult to find sets where more than five of the twenty-four possible permutations are English words; and for

$k = 5$, eight English words among the 120 permutations is an excellent score.

In general, $(k + 1)k! = (k + 1)!$, since the product of all the numbers from 1 through k, multiplied by $k + 1$, is the product of all the numbers from 1 through $k + 1$. This gives the relation

$$k! = \frac{(k + 1)!}{k + 1},$$

and if $k = 0$, $0! = \dfrac{1!}{1} = 1$, which is generally adopted as the definition of *zero factorial*. (The original definition did not cover $k = 0$, so that this extends it.)

A famous approximate formula for $k!$ called *Stirling's formula* is

$$k! \sim \sqrt{2\pi k} \left(\frac{k}{e}\right)^k.$$

Here π (pi) = 3.14159 . . . is the ratio of the circumference to the diameter of a circle, and e = 2.71828 . . . is another famous mathematical constant. (It is the base of "natural logarithms.") The sign $\sqrt{}$ is read as "the square root of"; and the sign $\sim$, which is read "is *asymptotic* to," means that the ratio of the two sides gets closer and closer to 1 as k gets larger and larger. For any specific choice of k, an *approximate* value for $k!$ is found, and the approximation becomes better and better, on a percentage basis, as larger and larger values of k are chosen. The derivation of Stirling's formula is completely outside the range of combinatorial analysis, and will not be given here. However, to get an approximate answer to the number of ways in which a deck of fifty-two cards can be dealt out, in Stirling's formula ($k = 52$)

$$52! \approx \sqrt{104\pi} \left(\frac{52}{e}\right)^{52} \approx 18.076 \, (19.130)^{52} \approx 8.053 \times 10^{67}$$

is obtained, where $\approx$ means "approximately equal to." For numbers that are so huge, it is frequently unimportant to know the exact digits, and an approximation is sufficient.

Suppose that a set of k distinct objects is given and that only r of them are to be selected; these will be labeled first, second, and so on, through rth. The number of distinct ways of selecting such an ordered subset of r objects from a set of k objects will be denoted by $(k)_r$.

Theorem 3.

$$(k)_r = \frac{k!}{(k-r)!} = k(k-1)(k-2)\ldots(k-r+1).$$

Proof. The first object may be selected in k ways, the second object in $k - 1$ ways, and in general the j-th object in $k - j + 1$ ways. In particular, the rth and final object can be selected in $k - r + 1$ ways; thus, there are $k(k-1)(k-2) \ldots (k-r+1)$ choices altogether. Since this is the product of the numbers from 1 to k *except* for the numbers from 1 to $k - r$, it equals $\dfrac{k!}{(k-r)!}$.

Exercises

3. In a certain contest, three distinct letters of the alphabet must be selected and written down in the right order. There is only one way to win. How many ways are there to lose?
4. Each of six players is dealt one card from a deck of fifty-two distinct cards. How many different situations (combinations of people and cards) may result?

When five cards are dealt to a player from a deck of fifty-two cards, he looks at his "hand" without regard to the *order* in which the cards were dealt. That is, the number of possible "poker hands" is the number of ways of selecting an unordered subset of five cards from the fifty-two-card deck. Similarly, the number of possible "bridge hands" is the number of ways of selecting an unordered subset of thirteen cards from the fifty-two-card deck. In general, if there is a set of k distinct objects, and r of them are to be selected, the number of possible unordered subsets of r objects from the set of k objects will be denoted by $\binom{k}{r}$.

Theorem 4.

$$\binom{k}{r} = \frac{k(k-1)(k-2)\ldots(k-r+1)}{r(r-1)(r-2)\ldots 2 \cdot 1} = \frac{(k)_r}{r!} = \frac{k!}{r!(k-r)!}.$$

Proof. The number of *ordered* subsets of r objects from a set of k objects is $(k)_r$. Given these r objects, they could be arranged in any of $r!$ distinct permuted orders and still constitute the same *unordered subset*. Thus, the number of unordered subsets is $\dfrac{(k)_r}{r!}$. From Theorem 3, this is also equal to

$$\frac{k!}{r!(k-r)!} \quad \text{and to} \quad \frac{k(k-1)(k-2)\ldots(k-r+1)}{r(r-1)(r-2)\ldots2\cdot1}.$$

By Theorem 4, the number of possible poker hands is

$$\binom{52}{5} = \frac{52\cdot51\cdot50\cdot49\cdot48}{5\cdot4\cdot3\cdot2\cdot1} = 2{,}598{,}960;$$

while the number of possible bridge hands is

$$\binom{52}{13} = \frac{52\cdot51\cdot50\cdot49\cdot48\cdot47\cdot46\cdot45\cdot44\cdot43\cdot42\cdot41\cdot40}{13\cdot12\cdot11\cdot10\cdot9\cdot8\cdot7\cdot6\cdot5\cdot4\cdot3\cdot2\cdot1}$$

$$= 635{,}013{,}559{,}600.$$

Exercises

5. In how many ways can four monominoes be placed on an 8 × 8 chessboard?
6. In how many ways can eight queens be placed on an 8 × 8 chessboard?
7. In how many ways can six pentominoes be chosen from the set of twelve distinct pentominoes?
8. A club wishes to hold meetings three evenings a week. How many possible choices are there for the nights on which meetings are to be held?

The *binomial theorem* of algebra is the formula

$$(x + y)^k = x^k + \binom{k}{1}x^{k-1}y + \binom{k}{2}x^{k-2}y^2 + \binom{k}{3}x^{k-3}y^3$$

$$+ \ldots + \binom{k}{k-1}xy^{k-1} + y^k.$$

For this reason, the numbers $\binom{k}{r}$ are known as the *binomial coefficients*; a coefficient is a multiplier, or factor, in mathematical operations. Often, $\binom{k}{r}$ is read as "the binomial coefficient k over r." The reader is encouraged to try to prove the binomial theorem from what he has already learned about binomial coefficients.

While the number of subsets containing *no* objects from a set of k objects has not been discussed, the convention that there is one such empty subset, which agrees with $\binom{k}{0} =$

$\dfrac{k!}{0!\,k!} = 1$ will be adopted. In the binomial theorem, the initial

term x^k may thus be regarded as having the coefficient $\binom{k}{0}$,

while the final term y^k has the coefficient $\binom{k}{k}$.

The seventeenth-century French mathematician Blaise Pascal arranged the binomial coefficients into the following triangle:

$$\binom{0}{0}$$

$$\binom{1}{0}\quad\binom{1}{1}$$

$$\binom{2}{0}\quad\binom{2}{1}\quad\binom{2}{2}$$

$$\binom{3}{0}\quad\binom{3}{1}\quad\binom{3}{2}\quad\binom{3}{3}$$

$$\binom{4}{0}\quad\binom{4}{1}\quad\binom{4}{2}\quad\binom{4}{3}\quad\binom{4}{4}$$

and so on.

This array is known as Pascal's triangle, and the explicit numerical values are shown below.

$$
\begin{array}{ccccccccc}
 & & & & 1 & & & & \\
 & & & 1 & & 1 & & & \\
 & & 1 & & 2 & & 1 & & \\
 & 1 & & 3 & & 3 & & 1 & \\
1 & & 4 & & 6 & & 4 & & 1 \\
\end{array}
$$

```
                1
              1   1
            1   2   1
          1   3   3   1
        1   4   6   4   1
      1   5  10  10   5   1
    1   6  15  20  15   6   1
  1   7  21  35  35  21   7   1
1   8  28  56  70  56  28   8   1
```

and so on.

Since the time of Pascal, literally hundreds of identities concerning this numerical triangle have been proved. Here are some for the reader to try.

1. The sums of the rows are the successive powers of 2. That is,

$$\binom{k}{0} + \binom{k}{1} + \binom{k}{2} + \ldots + \binom{k}{k} = 2^k.$$

2. The alternating sums of the rows (that is, the first term minus the second term plus the third term minus the fourth term, and so on) are all *zero* except for the top row, thus,

$$\binom{k}{0} - \binom{k}{1} + \binom{k}{2} - \binom{k}{3} + \ldots \pm \binom{k}{k} = 0.$$

3. Every entry is the sum of the two entries approximately above it (northeast and northwest of it, so to speak). That is,

$$\binom{k}{r} = \binom{k-1}{r-1} + \binom{k-1}{r}.$$

4. Each row is symmetric. That is, $\binom{k}{r} = \binom{k}{k-r}.$

5. The sum of the squares across any row equals the middle entry in the row twice as far down, excluding the top row. That is,

$$\binom{k}{0}^2 + \binom{k}{1}^2 + \binom{k}{2}^2 + \binom{k}{3}^2 + \ldots + \binom{k}{k}^2 = \binom{2k}{k}.$$

It is now possible to answer a question raised earlier: namely, what is the number of unordered combinations of n symbols from a basic alphabet of k symbols?

Theorem 5. The number of possible unordered combinations of n selections from a basic alphabet of k symbols is

$$\binom{k+n-1}{n} = \binom{k+n-1}{k-1}.$$

Proof. First, to illustrate the method of proof, a typical unordered combination of $n = 7$ selections from an alphabet of $k = 4$ symbols will be used, the symbols being a, b, c, d. If the combination is $c\ a\ d\ b\ a\ a\ b$, it is written in the "standardized form" $a\ a\ a|b\ b|c|d$. Since *unordered* combinations are being studied, the standardized form is the same unordered combination as the original. In general, an unordered combination is put into standardized form by arranging all the occurrences of the first symbol together at the beginning, then a vertical demarcation line, then all occurrences of the second symbol, and so on. If one of the four alphabet symbols fails to occur, the vertical demarcation line for it is nevertheless indicated. In this way, there are always n alphabetic symbols in the standardized form, and $k - 1$ vertical demarcation lines. With both alphabetic symbols and demarcation lines classified as *marks*, one gets a total of $n + k - 1$ marks. In the illustrative example, there are $7 + 4 - 1 = 10$ marks. The specification of *which* $k - 1$ of a string of $n + k - 1$ marks are to be demarcation lines is *equivalent* to writing down an unordered combination of n selections from a k-symbol alphabet in standardized form. For example, if in the sequence of ten marks, ----------, it is specified that the fourth, seventh, and ninth are to be demarcation lines, the result is ---|--|-|-, which is automatically filled in as $a\ a\ a|b\ b|c|d$. Demarcation lines specified at the first, fourth, and fifth places give |--||-----, which would correspond to $|b\ b||d\ d\ d\ d$, and similarly, every choice of locations for the demarcation lines is a *different* unordered combination in standard form. By Theorem 4, there

are exactly $\dbinom{n + k - 1}{k - 1}$ ways to select $k - 1$ of the $n + k$

$- 1$ marks to be the demarcation lines, and this is therefore the number of unordered combinations of n objects selected from a k-symbol alphabet. Finally,

$$\binom{n + k - 1}{k - 1} = \frac{(n+k-1)!}{(k-1)!\ n!} = \binom{n + k - 1}{n}.$$

Examples

1. The number of unordered combinations of five selections from the binary digits is $\dbinom{5 + 2 - 1}{5} = \dbinom{6}{5} =$

6. In standardized form, these are $0\ 0\ 0\ 0\ 0|,0\ 0\ 0\ 0|1,$ $0\ 0\ 0|1\ 1,\ 0\ 0|1\ 1\ 1,\ 0|1\ 1\ 1\ 1,$ and $|1\ 1\ 1\ 1\ 1$. Each of these unordered combinations corresponds to one or more ordered combinations, as shown below. Two ordered combinations correspond to the same unordered combination if they have the same number of 1's.

Unordered *Combination*	*Ordered Combinations*
00000	00000
00001	00001, 00010, 00100, 01000, 10000
00011	00011, 00101, 01001, 10001, 00110, 01010, 10010, 01100, 10100, 11000
00111	00111, 01011, 10011, 01101, 10101, 11001, 01110, 10110, 11010, 11100
01111	01111, 10111, 11011, 11101, 11110
11111	11111

Note that the sizes of these six categories of ordered combinations are the numbers 1, 5, 10, 10, 5, 1, which form a row of Pascal's triangle.

2. The number of unordered combinations of three

selections from a 3-symbol alphabet is $\binom{3+3-1}{3}=$

$\binom{5}{3}$ = 10. Specifically, they are $a\,a\,a$, $a\,a\,b$, $a\,a\,c$,

$a\,b\,b$, $a\,b\,c$, $a\,c\,c$, $b\,b\,b$, $b\,b\,c$, $b\,c\,c$, $c\,c\,c$.

Exercises

9. Determine the number of unordered combinations of four selections from a 5-symbol alphabet, and list them.

10. A mixed assortment of fruit, consisting of apples, oranges, and pears, is on sale for a dollar per dozen. How many distinct assortments might a customer get for his dollar?

11. Given, k symbols of r distinct types. There are k_1 of the first type, k_2 of the second type, k_3 of the third type, ..., and k_r of the r-th type. (Of course, $k_1 + k_2 + k_3 + \ldots + k_r = k$.) Show that the number of distinguishable permutations of these k symbols is

$$\frac{k!}{k_1!\,k_2!\,k_3!\,\ldots\,k_r!}$$

12. Use the result of the previous exercise to compute the number of distinct permutations of PEPPER and of MISSISSIPPI.

INCLUSION AND EXCLUSION

There is an ancient "folk theorem" asserting that a horse has at least twelve legs, because it has two legs in front, two in back, two on each side, one at each corner, and that doesn't even take into account the ones at the bottom!

A similar "paradox" is designed to prove that there are no workdays in the business year. Of the 365 days, 104 are weekends, 7 are paid holidays, and 10 are paid vacation. Moreover, one-third of the day, or 122 days a year, is spent sleeping, and still another one-third of the day (another 122 days a year) lies outside the 8-hour workday. This gives $104 + 7 + 10 + 122 + 122 = 365$ days a year *not* devoted to work.

It is obvious that the resolution of these paradoxes involves the fact that there are "overlaps" in the categories considered for which proper accounting must be made. Thus, some of the horse's front legs may be on the right side and some of the corner legs may also be on the bottom. Similarly, some of the eight hours a day spent sleeping occur on weekends and holidays and should not be subtracted twice from the 365-day year. What is remarkable is that there is a very simple formula, known as the "principle of inclusion and exclusion" (and also known as the "principle of cross-classification") that gives the correct answers to problems of this type by taking into precise account the amount of overlap.

Theorem 6. In a set of N objects, suppose N_1 of them have a property P_1, N_2 of them have a property P_2, and so on, and N_r of them have a property P_r. Then the number of objects having *none* of the properties $P_1, P_2, \ldots, P_r$ is N_0, given by

$$N_0 = N - (N_1 + N_2 + \ldots + N_r) + (N_{1,2} + N_{1,3} + \ldots + N_{2,3}$$

$$+ \ldots + N_{r-1,r}) - (N_{1,2,3} + N_{1,2,4} + \ldots + N_{r-2,r-1,r})$$

$$+ - \ldots \pm (N_{1,2,3\ldots r}),$$

where $N_{i,j,\ldots,m}$ is the number of objects having all the properties $P_i, P_j, \ldots, P_m$.

Illustration. Let $N = 365$ days, and let $P_1 = $ weekends, $P_2 = $ paid holidays, $P_3 = $ paid vacation, $P_4 = $ sleep time, and $P_5 = $ the time that is spent neither working nor sleeping. Then, in days, $N_1 = 104$, $N_2 = 7$, $N_3 = 10$, $N_4 = 122$, and $N_5 = 122$. However, $N_{1,4} = \frac{104}{3}$, $N_{2,4} = \frac{7}{3}$, $N_{3,4} = \frac{10}{3}$, $N_{1,5} = \frac{104}{3}$, $N_{2,5} = \frac{7}{3}$, and $N_{3,5} = \frac{10}{3}$. All the other $N_{i,j,\ldots,m}$ are zero. Hence, N_0, the time actually worked, is

$$N_0 = N - (N_1 + N_2 + N_3 + N_4 + N_5) + (N_{1,4} + N_{2,4}$$
$$+ N_{3,4} + N_{1,5} + N_{2,5} + N_{3,5})$$

$$= 365 - 365 + \frac{242}{3} = 80\tfrac{2}{3} \text{ days}$$

$$= 1{,}936 \text{ hours} = 48.4 \text{ 40-hour weeks.}$$

Proof. From the set of N objects, we subtract the N_1 that have property P_1, the N_2 that have property P_2, and so on. However, the objects that have two such properties have been subtracted *twice*, and must therefore be restored *once*, so that all terms $N_{i,j}$ are added back in. The objects that have three of the basic properties, say P_i, P_j, and P_k, have now been subtracted three times by $-N_i -N_j -N_k$, and added back three times by $+N_{i,j} +N_{i,k} +N_{j,k}$, and must therefore be subtracted out once more, so that all terms $N_{i,j,k}$ must be *subtracted*. In general, an object with t of the r properties will be subtracted out t times, added back $\binom{t}{2}$ times, subtracted again $\binom{t}{3}$ times, added again $\binom{t}{4}$ times, and so on, for a net contribution of

$$-t + \binom{t}{2} - \binom{t}{3} + \binom{t}{4} - + \ldots \pm \binom{t}{t},$$ which is one less

than $1 - \binom{t}{1} + \binom{t}{2} - \binom{t}{3} + - \ldots \pm \binom{t}{t} = 0$. That is, an

object with one or more of the specified r properties will be subtracted out exactly once.

Examples

1. A fisherman wishes to fish in the $1{,}000 \times 1{,}500$ mile area shown in figure 91. However, he must stay out of the three square test areas A, B, C, as shown, each of which is 500 miles on a side, and the centers of any two of which are 400 miles apart. How many square miles remain in which he can fish?

    Symbolically, the fishing area, F, is the total area, T, less the areas of A, B, and C, plus the areas of AB, AC, and BC (each of which includes ABC), minus the area ABC. The *numerical* values are left as an exercise for the reader.

Chapter 5

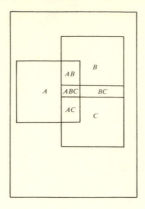

Figure 91.
The fisherman's
dilemma.

2. Suppose that the number n has the distinct prime factors $p_1, p_2, \ldots, p_r$, and the number of numbers from 1 to n having no prime factor in common with n are sought. This number is denoted $\phi(n)$, and is called *Euler's phi-function of n*. (For various reasons, the number 1 is *not* regarded as a prime number. If it were, every number up to n would have this factor in common with n, and $\phi(n)$ would always be 0.) Thus $\phi(1) = 1$, $\phi(2) = 1$, $\phi(3) = 2$, $\phi(4) = 2$, $\phi(5) = 4$, $\phi(6) = 2$, $\phi(7) = 6$, $\phi(8) = 4$, and so on. Let P_1 be the property of being divisible by p_1, let P_2 be the property of being divisible by p_2, and so forth. Then, applying Theorem 6.

$$\phi(n) = n - (n_1 + n_2 + \ldots + n_r) + (n_{1,2} + n_{1,3} + \ldots)$$

$$- (n_{1,2,3} + \ldots) \pm \ldots = n - \left(\frac{n}{p_1} + \frac{n}{p_2} + \ldots + \frac{n}{p_r}\right)$$

$$+ \left(\frac{n}{p_1 p_2} + \frac{n}{p_1 p_3} + \ldots + \frac{n}{p_{r-1} p_r}\right)$$

$$- \left(\frac{n}{p_1 p_2 p_3} + \frac{n}{p_1 p_2 p_4} + \ldots\right) + - \ldots \pm \frac{n}{p_1 p_2 \cdots p_r}$$

$$= n\left(1 - \frac{1}{p_1}\right)\left(1 - \frac{1}{p_2}\right) \ldots \left(1 - \frac{1}{p_r}\right).$$

Exercises

13. For each value of n from 1 to 15, list the numbers up to n that have no prime factors in common with n. (The prime numbers in this range are 2, 3, 5, 7, 11, 13). Compare the sizes of these lists with the formula

$$\phi(n) = n\left(1 - \frac{1}{p_1}\right)\left(1 - \frac{1}{p_2}\right) \ldots \left(1 - \frac{1}{p_r}\right) \text{ just derived.}$$

14. Let $p_1, p_2, \ldots, p_m$ be all the prime numbers up to $\sqrt{x}$. Show that the number of prime numbers up to x, denoted $\pi(x)$, is given by

$$\pi(x) = m - 1 + [x] - \left(\left[\frac{x}{p_1}\right] + \left[\frac{x}{p_2}\right] + \ldots + \left[\frac{x}{p_m}\right]\right)$$
$$+ \left(\left[\frac{x}{p_1 p_2}\right] + \left[\frac{x}{p_1 p_3}\right] + \ldots + \left[\frac{x}{p_{m-1} p_m}\right]\right)$$
$$- \ldots \pm \left[\frac{x}{p_1 p_2 \cdots p_m}\right],$$

where $\left[\dfrac{x}{p}\right]$ is the integer part of $\dfrac{x}{p}$, that is, $\dfrac{x}{p}$ rounded *downward* if it is not an exact integer (an integer is any whole number). (*Hints*: (*a*) If a number between 1 and x is not prime, one of its prime factors must be less than $\sqrt{x}$. Why? (*b*) The number of multiples of p up to x is $\left[\dfrac{x}{p}\right]$. Why? (*c*) It must be remembered to include the *first m* primes in $\pi(x)$; and to reject 1 as a non-prime.)

15. The prime numbers up to $\sqrt{100}$ are 2, 3, 5, 7. Use the formula of the previous problem to determine $\pi(100)$. Does the answer agree with a list of the primes up to 100?

16. Among a population of 50 stray dogs, 12 had distemper, 18 had fleas, 11 had worms, 6 had rabies, 5 had distemper and fleas, 3 had fleas and rabies, 4 had distemper and worms, 5 had fleas and worms, and 2 had distemper, fleas, and worms. Assuming there were no other ailments or combinations thereof, how many of the dogs were healthy?

COUNTING THE DISSIMILAR CASES

To count the number of cows in a field, it suffices to add together the number of legs and divide by 4. To find the number of automobiles in a parking lot, one could count the number of wheels and divide by 4. However, if the parking lot contains bicycles and wheelbarrows as well as automobiles, then the number of vehicles is no longer one-fourth the total number of wheels. In fact, the number of vehicles, V, is given by the formula

$$V = \tfrac{1}{4}(A + 2B + 4W)$$

where A is the number of automobile wheels, B is the number of bicycle wheels, and W is the number of wheelbarrow wheels.

Of course, $\frac{1}{4}A$ is the number of automobiles, $\frac{1}{2}B$ is the number of bicycles, and W is the number of wheelbarrows, and the formula for V is simply the sum of these three terms. This expression, as will be seen, illustrates a very general counting formula, in which the individual terms may not correspond so simply to the objects being counted.

Suppose that gloves come in three colors: red (R), white (W), and blue (B). A left glove and a right glove are chosen at random. How many distinguishable cases are there? Since there were three possibilities for the *left* glove, and three possibilities for the *right* glove, there are $3^2 = 9$ distinguishable pairs, namely *RR, RW, RB, WR, WW, WB, BR, BW, BB.*

Suppose that instead of gloves, socks are being selected from the same three colors. Now there is no longer a distinction between left and right. The first impulse is to divide the previous answer by 2. However, since $\frac{9}{2}$ is not a whole number, it is clear that the approach was too naive. The dissimilar pairs can be grouped two at a time:

RW with *WR*

RB with *BR*

BW with *WB.*

However, the *matched* pairs must still be counted individually:

RR

WW

BB.

Thus, there are six distinguishable pairs of socks possible. A *formula* that gives the number, N, of distinguishable cases, after allowing interchange of left and right, where the original number of cases (not allowing for the symmetry) was T, is

$$N = \tfrac{1}{2}(T + C)$$

where C is the number of *symmetric* cases. In the case just considered, $T = 9$ and $C = 3$, so that $N = \frac{1}{2}(9 + 3) = 6$. More generally, if the socks came in a distinct colors, we would have $T = a^2$ and $C = a$, so that the number of distinguishable pairs of socks would be $N = \frac{1}{2}(a^2 + a) = \frac{1}{2}a(a + 1)$.

Interchanging left and right is an operation that, when repeated, returns the case to its original position. Any symmetry operation which is one-half of a two-step operation that returns to the starting point is called an *involution*. Some examples of involutions are: rotating a plane figure 180 degrees; turning something inside out; interchanging top and bottom; taking the reciprocal, $\dfrac{1}{x}$, of a nonzero number, x; and taking the negative,

$-x$, of any number, x. The formula $N = \frac{1}{2}(T + C)$ gives the number of distinguishable cases for any involution, where, as before, T is the total number of cases before considering the involution, and C is the number of cases whose appearance remains unchanged, or unaffected, or, to use the terminology of higher mathematics, *invariant* under the involution.

Exercises

17. Two three-digit numbers will be put in the same category if one of them contains the same digits as the other in reverse order. (Some numbers, like 131, will be in a category by themselves.) Into how many categories will the 1,000 numbers from 000 to 999 be placed?

18. All possible "words" consisting of any four English letters are formed. If words read backward are considered equivalent to words read forward, how many inequivalent four-letter "words" are there?

19. A string of beads is formed by placing k beads on the string, where each bead is any one of n colors. If a string is "turned around," it is not possible to tell that it was not originally constructed in the reverse direction. How many *distinguishable* strings of beads are there? (*Hint*: Treat the cases where k is even and the cases where k is odd separately.) Verify the results for $n = 2$ and 3 with $k = 4$ and 5.

20. Each card in a deck of blank index cards is marked with a number on both sides. First, a number from 1 to 5 is stamped on one side. Then a number from 3 to 9 is stamped on the reverse side. How many distinguishable situations can result? There is no way to distinguish "top" from "bottom" on a blank card. (*Hint*: It is easy to list all the distinct cases to check your result.)

21. A set of "double-6" dominoes contains every distinct pair of numbers from 00 to 66 (using the digits 0 through 6) exactly once. How many dominoes are there in a set? A set of double-9 dominoes contains every distinct pair from 00 to 99. How large is the double-9 set?

Suppose there is a deck of 3 inch × 5 inch blank index cards, and corners are to be cut out of some of the cards. (Imagine a cut corner as the removal of an isosceles right triangle whose leg is one-half inch long.) Any number of corners, from 0 to 4, may be clipped on a single card. How many distinguishable cases of mutilated cards are there?

If the *symmetries* of the cards did not have to be considered, the problem would be relatively easy. Specifically, if front could be distinguished from back, right from left, and top from bottom on each card, the corners would be designated upper right, upper left, lower right, and lower left. Each of these four corners could be either notched or unnotched, so that by Theorem 1, there are $2^4 = 16$ possibilities. However, with the four corners indistinguishable, another approach must be used.

The "brute force" approach is to enumerate cases:

1. No corners clipped.
2. One corner clipped.
3. Two adjacent corners on short side clipped.
4. Two adjacent corners on long side clipped.
5. Two diagonally opposite corners clipped.
6. Any three corners clipped.
7. All four corners clipped.

These seven conditions are shown in figure 92.

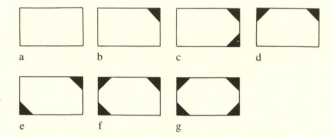

a   b   c   d

Figure 92. Seven distinguishable ways to clip the corners of a rectangular card.

e   f   g

A more methodical approach to the problem is to observe that *symmetries* of the index card correspond to three distinct *involutions*:

1. Rotating 180 degrees around the center (in the plane).
2. Rotating 180 degrees around the horizontal midline (turns card over).
3. Rotating 180 degrees around the vertical midline (turns card over).

(Strictly speaking, there is also a fourth symmetry operation, the "identity operator," which does not move the figure at all.)

Note that figures 92b and 92f are changed by all three of the involutions, while 92a and 92g are not changed by any of them. Also, 92c is left fixed by Point 2 above, 92d is left fixed by Point 3, and 92e is left fixed by Point 1.

To count the distinguishable cases in figure 92, one begins as before by dividing the original number of cases ($2^4 = 16$) by the number of symmetry operators (4, including the identity), but one must be careful to add in the terms for the number of cards that are symmetric under each of the involutions. The expression now becomes

$$N = \tfrac{1}{4}(T + C_a + C_b + C_c),$$

where N is the number of distinguishable cases, T is the total number of cases before allowing symmetries, C_a is the number of original cases left fixed by the Symmetry Operator 1, C_b is the number of original cases left fixed by Operator 2, and C_c is the number of original cases left fixed by Operator 3.

In the problem under consideration, $T = 16$. The cases, C_a, that are unaffected by 180-degree rotation around the center are those where the upper-right corner is the same as the lower left while the upper-left corner is the same as the lower right. There are clearly four such cases—the upper right–lower left pair may be either clipped or unclipped and the upper left–lower right pair may be either clipped or unclipped. Similarly, the number of cases for C_b is four, since again there are two *pairs* of corners: upper right–lower right, and upper left–lower left. Finally, there are also four cases for C_c, where the pairing now is upper corners together and lower ones together. The formula therefore yields

$$N = \tfrac{1}{4}(16 + 4 + 4 + 4) = 7,$$

as required.

Instead of *clipping* the corners of the index cards, they could be *dipped* into different colors of paint. Suppose there are three colors: red, white, and blue (where "white" could be the same as "undipped"). The total number of situations before considering symmetries is now $T = 3^4 = 81$, while $C_a = C_b = C_c = 3^2 = 9$. Thus, in this case, $N = \tfrac{1}{4}(81 + 9 + 9 + 9) = 27$.

Exercises

22. Draw the twenty-seven distinguishable cases involving red, white, and blue corners.
23. What is the formula for the number of distinguishable index cards when there are n possible colors for the corners?

To return briefly to the case of the vehicles in the parking lot, it will be agreed to call two wheels "equivalent" if they are on the same vehicle. Thus, by definition, the number of classes of equivalent wheels equals the number of vehicles. The wheelbase of an automobile is a rectangle, like an index card, and the symmetries of the index cards can be used not only for the automobiles, but also for the bicycles and the

Chapter 5

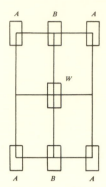

Figure 93. Wheels
of an automobile
(A), bicycle (B),
and
wheelbarrow (W).

wheelbarrows, as indicated in figure 93. In this context, $T = A + B + W$, the total number of wheels. Next, $C_a = W$, since rotating an automobile or a bicycle 180 degrees leaves none of the wheels in the same place, while the wheelbarrow wheel *does* remain fixed under these circumstances. Similarly, $C_b = W$, since again front and back wheels are being interchanged (though not in criss-cross fashion as before). Finally, $C_c = B + W$, since the right-to-left symmetry leaves bicycle wheels as well as wheelbarrow wheels fixed. Thus, as before, the number of *vehicles* is

$$N = \tfrac{1}{4}\{(A + B + W) + W + W + (B + W)\}$$

$$= \tfrac{1}{4}(A + 2B + 4W).$$

Exercises

24. A 2×3 hexomino is to be made of six square tiles. Each tile can be chosen from a set of five colors. How many distinguishable colored hexominoes can be formed?

25. What capital letters in the English alphabet have the same set of four symmetries as the index card (and other rectangles)? How would you approximate these shapes as polyominoes?

26. An $a \times b$ rectangle (where a and b are unequal) is to be pasted together using black and white 1×1 squares. How many distinguishable patchwork rectangles can result? Colors are visible on both sides, and turning over is permitted. (*Hint*: Distinguish the cases where a and b are both odd, both even, and of unlike parity.)

The squares of a 2×2 checkerboard are to be colored in one of two colors. How many distinguishable cases are there if *rotations* of the checkerboard, but not *reflections* are allowed as symmetries? The six distinguishable cases are shown in figure 94. The number of cases would not be reduced fur-

Figure 94.
Distinguishable
ways to color a
2 × 2 checkerboard
with 2 colors.

ther by allowing reflections as symmetries, too. However, if the same problem is considered for the 3×3 checkerboard, there are examples of mirror twins that are distinct if only rotations are allowed. Two such examples are shown in figure 95.

 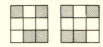

Figure 95. Distinct mirror-image pairs.

The *symmetries* of a square under rotations are 0-degree rotation (the identity), 90-degree rotation, 180-degree rotation, and 270-degree rotation (or, equivalently, −90-degree rotation). To count the number, N, of distinguishable cases under these rotations, the same sort of formula as before is used:

$$N = \tfrac{1}{4} (T + C_{90} + C_{180} + C_{270}),$$

where C_θ is the number of cases left fixed by a rotation of angle θ. Note that the total number, T, of cases *before* considering rotations is really C_0; and in general, the T in these formulas is the number of cases left fixed by the "identity operator." Also, it will always turn out that $C_{90} = C_{270}$, because if a figure looks the same after it is rotated *forward* by 90 degrees, it will certainly look the same if it is rotated *backward* by 90 degrees. Thus, our formula can be rewritten as

$$N = \tfrac{1}{4} (T + 2C_{90} + C_{180}).$$

For the 2×2 checkerboard in 2 colors, $T = 2^4 = 16$, since each of 4 squares can be colored either light or dark. Next, $C_{90} = 2$, since the 2×2 board will only look the same after 90-degree rotation if all squares are the same color—either all light or all dark. Finally, $C_{180} = 2^2 = 4$, since the requirement for invariance under 180-degree rotation is that opposite corners must have the same color, with two possible colors for each of two pairs of corners. The formula therefore gives

$$N = \tfrac{1}{4} (16 + 2 \cdot 2 + 4) = 6,$$

which is the number of cases shown in Figure 94.

Exercises

27. How many distinguishable 2×2 boards, under rotation, are there if each square may be one of three colors? Check your answer by drawing the distinguishable cases.
28. A square is cut into four triangular regions by its diagonals. If three colors are allowed for the triangular regions, how many distinguishable colorings, allowing rotation, are there? (Is this the same as the answer to the previous problem?) Arrange a complete set of these colored squares in a rectangle, so that adjacent triangular regions of adjacent squares have the same color and the outer border is colored in a single color.

29. How many distinguishable 2×2 boards, allowing rotation and not reflection, are there if four colors may be used for the squares? What is the general answer if *n* colors are allowed? Can you show directly (algebraically) that this number is always an integer?
30. How many 3×3 boards, distinguishable under rotation, are there if two colors may be used for the squares? Can you draw all these cases?
31. How many distinguishable 3×3 boards are there, under rotation, if *n* colors may be used for the squares?
32. Derive general formulas for the number of distinguishable $k \times k$ boards, under rotation, where *n* colors may be used for the squares. (*Hint*: Treat odd *k* and even *k* as separate cases.)
33. Find an octomino that is symmetric under 90-degree rotation. In how many ways could eight square tiles, each either black or white, be glued together to form this figure?
34. Beads are available in *n* colors. Four beads are chosen and placed on a string, and the ends of the string are tied together. How many of the resulting necklaces are distinguishable if two necklaces that differ only by rotation of the beads around the string are regarded as being the same?

It is now possible to consider the full set of symmetries of the square, including rotations and reflections. Consider the square in figure 96. In addition to the four rotations of 0, 90, 180, and 270 degrees around the center, *C*, there are reflections around the four axes: horizontal axis *HH*, vertical axis *VV*, diagonal axis $D_1 D_1$, and diagonal axis $D_2 D_2$. (These eight symmetries comprise what is known as the *dihedral group of the square*.)

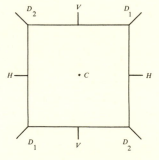

Figure 96. The 8 symmetries of the square.

As a typical problem, suppose one monomino (or one checker) is to be placed on the 8×8 board. It is clear from

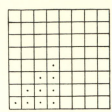

Figure 97. The 10 inequivalent locations
for a monomino on the checkerboard.

figure 97 that there are ten *inequivalent* locations under rotation and reflection. However, the answer also can be *computed* by a formula for the number N of inequivalent cases under all the (dihedral) symmetries of the square. Analogous to the previous cases, the formula is

$$N = \frac{1}{8}(T + C_{90} + C_{180} + C_{270} + C_V + C_H + C_{D_1} + C_{D_2})$$
$$= \frac{1}{8}(T + 2C_{90} + C_{180} + 2C_H + 2C_D).$$

Here, T is the total number of cases before symmetry considerations begin, and C_{90}, C_{180}, C_{270} are as in the previous formula. Also, C_V, C_H, C_{D_1}, and C_{D_2} are the number of configurations left invariant by reflections in the vertical, horizontal, first diagonal, and second diagonal axes, respectively. As before, $C_{90} = C_{270}$, and moreover $C_V = C_H$ and $C_{D_1} = C_{D_2}$, which leads to the second form of the formula.

For the 1-monomino problem at hand, $T = 64$, since there are initially 64 possible locations for a monomino on an 8 × 8 board. Next, $C_{90} = 0$, since a 90-degree rotation will surely *move* the monomino to a previously empty square. Likewise, $C_{180} = 0$, and $C_H = 0$. However, $C_D = 8$, since the monomino could be placed anywhere along a diagonal and not have its position affected by reflection in that diagonal. Thus,

$$N = \frac{1}{8}(T + 2C_D) = \frac{1}{8}(64 + 16) = 10, \text{ as required.}$$

As another example, suppose two monominoes are to be placed on a 4 × 4 board. In this case, $T = \binom{16}{2} = 120$.

Again, $C_{90} = 0$; but $C_{180} = 8$, since for any location of one monomino in the bottom half, there is a corresponding location, rotating 180 degrees around the center, in the top half.

Similarly, $C_H = 8$. Finally, $C_D = 6 + \binom{4}{2} = 12$, since there are six ways to place one monomino above the diagonal, using the other as its reflection *below* the diagonal, and $\binom{4}{2}$ ways to place both monominoes on the diagonal. This leads to the following number of inequivalent cases:

$$N = \tfrac{1}{8}(T + C_{180} + 2C_H + 2C_D)$$

$$= \tfrac{1}{8}(120 + 8 + 16 + 24) = 21.$$

The 21 cases are shown in figure 98.

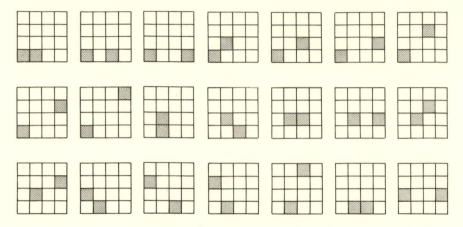

Figure 98. The 21 inequivalent ways to place 2 monominoes on a 4 × 4 board.

Exercises

35. In how many inequivalent ways can three monominoes be placed on a 3 × 3 board? (Compare the formula with a drawing of all cases.)

36. In how many inequivalent ways can four monominoes be placed on the 4 × 4 board? On the 6 × 6 board? On the 8 × 8 board?

37. In how many inequivalent ways (under the dihedral symmetry group) can a 4 × 4 checkerboard be colored in three colors?

38. In how many inequivalent ways can a 3 × 3 tic-tac-toe board be covered with five X's and four O's?

39. In how many ways can six monominoes be placed on an 8 × 8 checkerboard so that every rank and every file contains an *even* number of monominoes? First determine the number, T, of solutions *without* regard to symmetries, and then determine the number N of solutions inequivalent under the rotations and reflections of the square.

40. In how many inequivalent ways can one monomino be

placed on an $n \times n$ board? How many inequivalent ways are there for two monominoes? (*Hint*: It is often helpful to separate the even from the odd values of *n*.)

41. Classify the twelve pentominoes according to the type of symmetry groups they possess. Which one has the same symmetries as the square? Which have an involution as a symmetry? Which have only the trivial identity symmetry?

42. In how many distinguishable ways can be the five squares of the X pentomino be colored using three possible colors? How many of these actually make use of all three of the colors?

43. In how many ways can four rooks be placed on a 4 × 4 board so that no two can attack one another? Solve the problem first without regard for symmetries and then determine the number of cases distinguishable under the rotations and reflections of the square.

44. Solve the preceding problem for six rooks on a 6 × 6 board and for eight rooks on an 8 × 8 board.

THE PÓLYA-BURNSIDE FORMULA

The enumeration formulas of the previous section are all special cases of a very general formula attributed to William Burnside, although he was not the first one to discover it, and applied with great success by the Hungarian-American mathematician G. Pólya in the 1940's to problems of mathematical and scientific interest. The general formula may be stated as follows:

Theorem 7. Let *S* be any finite collection of objects, and let *G* be a finite group of *symmetries* for these objects, with *n* symmetry operations $g_1, g_2, \ldots, g_n$ comprising *G*. (One of these symmetry operations must be the identity operator.) Let $C(g)$ denote the number of objects in the collection, *S*, left fixed by the symmetry, *g*, of *G*. Then the number, *N*, of objects in *S distinguishable relative to the symmetries of G* is given by

$$N = \frac{1}{n} [C(g_1) + C(g_2) + C(g_3) + \ldots + C(g_n)].$$

(If g_1 is the identity operator of *G*, then $C(g_1) = T$, the total number of objects in *S*.)

While this formula is simple in appearance and relatively easy to apply, a rigorous proof of it would require an excursion into advanced mathematical theory. The interested reader .

is referred instead to the author's article on discrete classification and to John Riordan's volume, both cited in the Bibliography, for a complete proof and more extensive applications.

The reader should attempt to derive each of the four formulas of the previous section (for involution, rectangle group,
rotation group of the square, and dihedral group of the square)
from the general formula of Theorem 7.

For every regular *polygon* (a polygon is a closed plane figure bounded by straight lines), there is the group of rotations
(called the *cyclic* group of the polygon, because it consists of
all the rotations by multiples of $360/r$ degrees, which have
the effect of cycling the polygon about its center), and the
group of rotations and reflections (called the *dihedral* group
of the polygon). For a regular r-gon, the cyclic group consists
of r symmetries, and the dihedral group consists of $2r$ symmetries.

Solid figures also have interesting symmetry groups. For
example, the group of all spatial rotations of the regular *tet-
rahedron* (a solid of four faces) is one of twelve symmetries.
The rotation group of the cube, and also of the regular *octa-
hedron* (a solid of eight faces), is a group of twenty-four symmetries. Moreover, the rotation group of the regular *dodeca-
hedron*, a solid with twelve regular pentagons as faces, and
also of the regular *icosahedron*, which has twenty equilateral
triangles as faces, is a group of sixty symmetries.

For the ambitious reader, the adaptation and application of
the Pólya-Burnside formula of Theorem 7 to some of these
new situations is developed in the following final exercises of
the chapter.

Exercises

45. The six edges of a regular hexagon are each to be drawn
    in either black or red. How many cyclically distinct
    hexagons can result? (*Hint*: $N = \frac{1}{6} (C_0 + C_{60} + C_{120} +
    C_{180} + C_{240} + C_{300})$.) Draw all the cases.

46. In the previous problem, suppose it is also permitted to
    turn the hexagons over. How many distinguishable cases
    now exist?

47. Rework the preceding two problems allowing three colors
    for the edges. Then allow four colors. Finally, examine
    the problem when k colors are used.

48. The vertices of an equilateral triangle are to be colored
    from a set of five colors. If rotations and reflections of
    the triangle are permitted, how many distinct cases can
    result?

49. Let p be any prime number except 2 (thus, $p = 3, 5, 7, 11, 13, 17, \ldots$). Beads are available in b colors, and p beads are put on a string. How many distinct strings can be formed? The ends of the string are tied together to form a necklace. Under cyclic rotations, how many distinct necklaces are there? If it is also allowed to turn the necklaces over, how many distinguishable cases are there? Verify your result for $b = 2$ colors, and $p = 5$ beads on a string; also for $b = 4$, $p = 3$.

50. Identify the twenty-four rotational symmetries of a cube. (*Hint*: There are three kinds of *axes* going through the center of a cube—face-to-face, edge-to-edge, and vertex-to-vertex. Examine the kinds of rotations around each such axis.)

51. Six distinct colors of paint are available, and the six faces of a cube are to be painted, each with a different color. Show that this can be done in thirty essentially distinct ways.

52. How many distinct cubes are there with three black faces and three white faces?

53. How many distinct cubes are there with two red, two white, and two blue faces?

54. In how many essentially different ways can the eight *vertices* of a cube each be either labeled or unlabeled? How many of these cases involve four labeled vertices and four unlabeled vertices? Draw all of these 4-and-4 cases.

6 Bigger Polyominoes and Higher Dimensions

THE MATHEMATICAL theory of enumeration and certain notions of symmetry, including rotations and reflections, were developed in some detail in the last chapter. Some of these ideas will now be applied to the further study of polyominoes and to their extensions in more than two dimensions.

ONE-SIDED POLYOMINOES

Up to this point it has been assumed that polyominoes can be both rotated and reflected, or flopped over, at will. However, if they are confined strictly to the plane, the figures can still be rotated, but cannot be reflected. Polyominoes that cannot be turned over may be termed one-sided, and, in general, eliminating reflection increases the number of *n*-ominoes. Thus, although there are still only one monomino, one domino, and two trominoes, seven tetrominoes are now distinguished (instead of five), eighteen pentominoes (instead of twelve), sixty hexominoes (instead of thirty-five), and so on. As with the five two-sided tetrominoes, the seven one-sided tetrominoes cannot be arranged in any rectangle, but are shown in a connected pattern in figure 99. The eighteen one-sided pentominoes are arranged in a 9 × 10 rectangle in figure 100. (The percentage of one-sided *n*-ominoes that are distinct from their mirror images increases toward 100 percent as *n* increases.) Now, although the possibility of considering one-sided polyominoes has been pointed out, attention will again be focused on the two-sided case; but it should be remembered that this distinction can always be made.

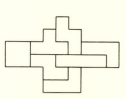

Figure 99. An almost symmetric pattern with the one-sided tetrominoes.

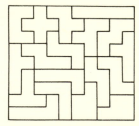

Figure 100. The 18 one-sided pentominoes arranged in a rectangle.

Hexominoes and Heptominoes

There are thirty-five distinct hexominoes, which cover an aggregate area of 210 squares. Rather surprisingly, however, they will not fit together into any *rectangle*, whether 3 × 70, or 5 × 42, or 6 × 35, or 7 × 30, or 10 × 21, or 14 × 15. (This was proved at the end of chapter 1 by a simple argument involving checkerboard coloring.) However, as shown some years ago in the British publication, no longer issued, called *Fairy Chess Review*, other interesting arrangements of the thirty-five hexominoes are possible. Two of these are shown in figures 101 and 102. (W. Stead found many of these.)

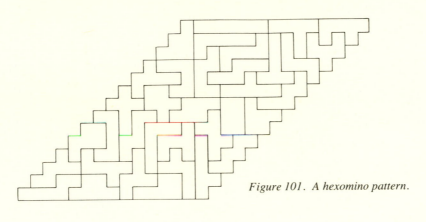

Figure 101. A hexomino pattern.

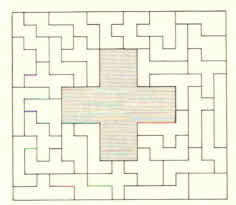

Figure 102. A symmetrical hexomino pattern.

There are 108 heptominoes, including one with a hole (shaded in figure 103). However, since this figure cannot be made to "fit" with other heptominoes except in patterns with "holes," one may sometimes prefer to exclude it from the approved list. The remaining heptominoes cover a total area of 7 × 107 = 749 squares. The only rectangle they can possibly occupy is

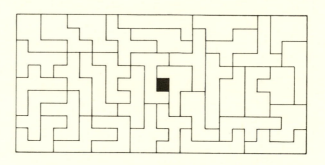

Figure 103.
Three congruent
rectangles made
with 108 heptomi-
noes.

a 7 × 107 one. A particularly interesting solution to fitting
the 107 heptominoes together, consisting of one 7 × 7 square
and four 7 × 25 rectangles, was found by David A. Klarner,
while a student at Humboldt State College in California, and
is shown in figure 104. The reader is invited to look for other
solutions to this type of construction. Another problem, also
first solved by Klarner, is to arrange all 108 heptominoes into
three 11 × 23 rectangles, each with the center square removed. Klarner's solution to this one appears in figure 103.

The Higher n-Ominoes

Table 3 was the best available in 1965 for the number of *n*-ominoes as a function of *n*. No reliable compilation of the number of *n*-ominoes had been reported for values of *n* greater than 10 by that time. See, however, Appendix D for the current (1993) status.

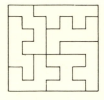

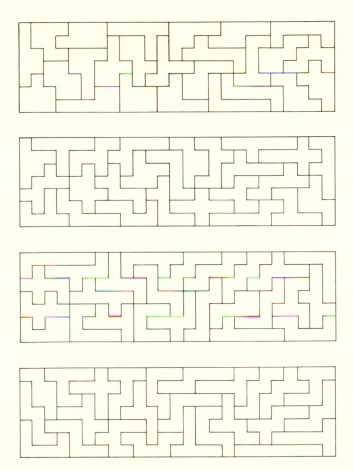

Figure 104. Four rectangles and a square with 107 heptominoes.

A multiply connected n-omino is one that has one or more holes in it. For example, an octomino that consists of the eight squares surrounding an empty square is multiply connected.

TABLE 3

	n									
	1	2	3	4	5	6	7	8	9	10
Number of simply connected n-ominoes	1	1	2	5	12	35	107	363	1248	4460
Number of multiply connected n-ominoes	0	0	0	0	0	0	1	6	37	195

The values for 9 and 10 come from an article in the *Canadian Journal of Mathematics* by R. C. Read, professor of mathematics at the University of the West Indies, Kingston, Jamaica. In the article, "Contributions to the Cell Growth Problem," methods are developed for finding the number of n-ominoes that fit inside rectangles of $1 \times p$, $2 \times p$, $3 \times p$, where p is any positive integer. Thus the final enumeration of 1285 *nonominoes* (nine squares) and 4655 *dekominoes* (ten squares) are the results of the tabulations shown below. Read's

TABLE 4

Nonominoes		Dekominoes	
Size of Rectangle	Number	Size of Rectangle	Number
1×9	1	1×10	1
2×8	7	2×9	9
2×7	28	2×8	40
2×6	22	2×7	52
2×5	3	2×6	15
3×7	49	2×5	1
3×6	188	3×8	63
3×5	210	3×7	332
3×4	42	3×6	550
3×3	1	3×5	255
4×6	97	3×4	21
4×5	383	4×7	155
4×4	181	4×6	822
5×5	73	4×5	1,304
TOTAL	1285	4×4	266
		5×6	240
		5×5	529
		TOTAL	4,655

enumeration counts simply connected and multiply connected n-ominoes together, so that $1285 = 1248 + 37$ and $4655 = 4460 + 195$, as noted in table 3.

An independent enumeration by the mathematician Thomas R. Parkin at the Aerospace Corporation, El Segundo, California, agrees with Read's published values except for the number of dekominoes in the 5×5 square, where Parkin correctly counts 529 instead of Read's originally published value of 340. (See table 4.)

Quite unrelated to Read's enumerative techniques is a systematic method for obtaining all the $(n + 1)$-ominoes from the n-ominoes. As the method is described, it will be illustrated by the case $n = 4$. First, all the n-ominoes are arranged according to the length of the longest straight row of squares they contain. For the tetrominoes, column I of figure 105 is obtained.

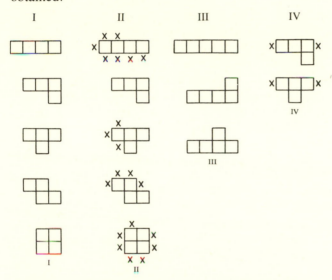

Figure 105. Generating the pentominoes from the tetrominoes.

If an n-omino possesses symmetries, its edges are closed off with x's, with the exception of one "typical region," as in column II. Beginning with the longest n-omino, an extra square may be attached to any *open* edge to obtain an $(n + 1)$-omino as in column III, while the next-longest n-ominoes are prevented from being made as long as the longest as in column IV. The first of those in column IV may now have a square added to any open edge as shown in figure 106. The n-omino just extended must now be prevented from occurring in the possible extensions of the other n-ominoes of the same length (see fig. 107a).

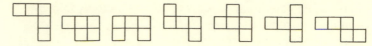

Figure 106. Pentominoes generated from the first entry of Column IV of figure 105.

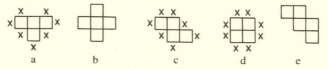

a   b   c   d   e

Figure 107. The last 2 pentominoes are found.

The next *n*-omino of the second-longest class is then extended in all remaining possible ways, as in figure 107b. After all the second-longest *n*-ominoes have been extended, the third-longest *n*-ominoes are prevented from being extended to the length of the second-longest ones (see figs. 107c and 107d), and the procedure is repeated, yielding figure 107e.

Proceeding in this fashion, all the $(n + 1)$-ominoes are obtained from the *n*-ominoes, each exactly once. (All twelve pentominoes, in fact, have now been obtained.) This method, while moderately lengthy, is another that is well suited to the way in which a computer is programmed—it is both deterministic and iterative. However, it is important to mention that there is one more rule required, in general, to make the transition from *n*-ominoes to $(n + 1)$-ominoes, for certain values of *n*, in order to guarantee the uniqueness of the $(n + 1)$-ominoes produced. For example, when $n = 5$, the P pentomino does not have any symmetries, and, if taken first in the class of length-3 pentominoes, would be restricted only as shown at the left of figure 108. If this figure is then extended, the

*Figure 108.
Hexominoes generated from the P pentomino.*

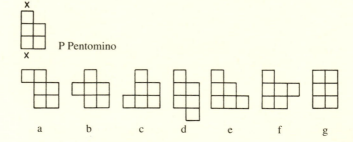

a   b   c   d   e   f   g

seven hexominoes of figure 108 result. But c and f are the same, and it would certainly be awkward to count the same polyomino twice in the attempt at enumeration. The duplication is due to the symmetry of 90-degree rotation in the following way. First, consider the polyominoes that are symmetric under 90-degree rotation, as shown in figure 109. An *n*-omino exceeding any of these figures either by one monomino, or by two monominoes 90 degrees apart, is capable of causing the same difficulty as encountered with the P pentomino. However, if the figure obtained from the 90-degree

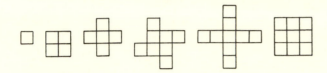

*Figure 109.
Polyominoes sym-
metric under
90-degree rotation.*

symmetric figure is itself symmetric (necessarily a reflectional symmetry) it does no actual harm. (That is, it contributes no duplicate to the enumeration.) From the symmetric polyominoes just shown in figure 109, the shapes exceeding them by one monomino and by two monominoes 90 degrees apart are as illustrated in figure 110. Of these, only the totally nonsymmetric ones (shaded in the figure) require special attention.

*Figure 110.
Higher-order poly-
ominoes derived
from figure 109.*

Compared to the literally thousands of polyominoes for n < 13 (that is, n *less than* 13), these twelve special cases are rare exceptions indeed. In a computer program or other exhaustive search, the extra $(n + 1)$-ominoes resulting from these cases could be left in the list of figures produced until the very end and then removed from the final compilation by the programmer.

The official count of the octominoes lists 369, including the six multiply connected cases of "doubtful" legitimacy shown in figure 111 where the shaded region indicates a hole. These are all extensions of the multiply connected heptomino shaded in figure 103.

Figure 111.
Multiply connected
octominoes.

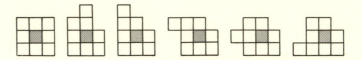

An exact expression for the number of n-ominoes, $P(n)$, in terms of n has not yet been found. As an empirical technique, a sequence C_n may be defined by

$$\frac{P(n + 1)}{nP(n)} = C_n.$$

and the sequence of values: $C_1 = 1$, $C_2 = 1$, $C_3 = .833$, $C_4 = .600$, $C_5 = .583$, $C_6 = .514$, $C_7 = .488$, $C_8 = .435$, $C_9 = .386$ may well converge to some limiting value. This would suggest an approximate expression of the sort

$$P(n) \approx k \cdot C^n \cdot n!,$$

where $C = \lim_{n \to \infty} C_n$. (That is, C is the limiting value of C_n as n tends toward infinity, ∞.)

In any case, it is easy to prove that

$$P(n + 1) < (2n + 1)P(n)$$

from the simple observation that there are at most $2n + 1$ places where an extra square can be attached to an n-omino, except for the straight n-omino. But, after allowing for symmetry, the straight n-omino can contribute at most $\frac{n + 3}{2}$, which is certainly less than $2n + 1$, to the list of $(n + 1)$-ominoes. This leads to the inequality $P(n) < \frac{(2n)!}{2^n n!}$, where $\frac{(2n)!}{2^n n!}$ grows like $\dfrac{1}{\sqrt{\pi n}}\,2^{n}n!$, which gives an upper bound on the size of

$P(n)$ that has a reasonably simple expression. This is certainly not the tightest possible result, but suggests a promising line of attack for the problem. (For an update, as of 1993, on the rate of growth of $P(n)$, and the exact value of $P(n)$ for all n up to 24, see Appendix D.)

SOLID POLYOMINOES

Numerous pentomino fans have suggested the possibility of gluing the twelve pentominoes together using five cubes instead of five squares each, and making solid patterns from them. More generally, one can list and enumerate the solid polyominoes of every number of sides. The solid polyominoes, allowing the pieces to be reflected and rotated, up through the tetrominoes, are shown in figure 112.

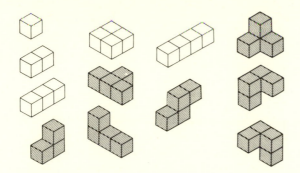

Figure 112. The lower-order solid polyominoes, with the Soma Cubes shaded.

Note that the first *nonplanar* figures (that is, those in which all the cubes do not lie in one plane) occur among the tetrominoes, since any three points lie in a plane, whereas four points (the centers of the four cubes) need not. Two of the three nonplanar tetrominoes are mirror twins, differing from each other as a right shoe differs from a left one.

A game invented by the contemporary Danish puzzle expert and game innovator Piet Hein and available commercially as the "Soma Cube" puzzle involves the seven solid polyominoes that are not simple rectangular solids (six tetrominoes and one tromino) shaded in figure 112. The object is to assemble these either into a $3 \times 3 \times 3$ cube, or into any of many other intriguing shapes. (A lengthy account of these problems is contained in the "Mathematical Games" column of *Scientific American* for September 1958.)

If one uses the twelve planar pentominoes in their solid form, a natural problem is to fit them into a 3 × 4 × 5 rectangular solid. One of the many solutions to this problem is shown with its step by step construction in figure 113.

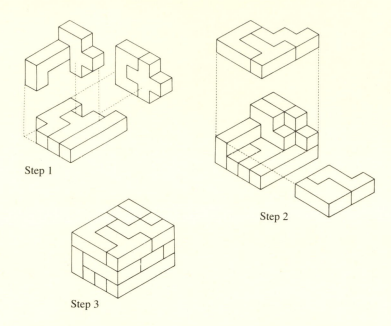

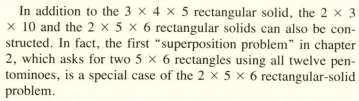

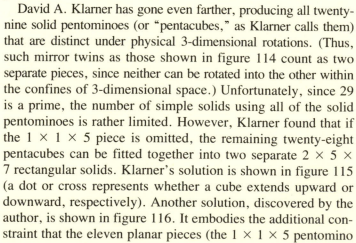

Step 1

Step 2

Step 3

Figure 113. Construction of a 3 × 4 × 5 solid using the 12 solid pentominoes.

Figure 114. Mirror-twin solid pentominoes.

In addition to the 3 × 4 × 5 rectangular solid, the 2 × 3 × 10 and the 2 × 5 × 6 rectangular solids can also be constructed. In fact, the first "superposition problem" in chapter 2, which asks for two 5 × 6 rectangles using all twelve pentominoes, is a special case of the 2 × 5 × 6 rectangular-solid problem.

David A. Klarner has gone even farther, producing all twenty-nine solid pentominoes (or "pentacubes," as Klarner calls them) that are distinct under physical 3-dimensional rotations. (Thus, such mirror twins as those shown in figure 114 count as two separate pieces, since neither can be rotated into the other within the confines of 3-dimensional space.) Unfortunately, since 29 is a prime, the number of simple solids using all of the solid pentominoes is rather limited. However, Klarner found that if the 1 × 1 × 5 piece is omitted, the remaining twenty-eight pentacubes can be fitted together into two separate 2 × 5 × 7 rectangular solids. Klarner's solution is shown in figure 115 (a dot or cross represents whether a cube extends upward or downward, respectively). Another solution, discovered by the author, is shown in figure 116. It embodies the additional constraint that the eleven planar pieces (the 1 × 1 × 5 pentomino having been discarded) all lie in the same 2 × 5 × 7 rectangular solid.

The reader is invited to look for additional simple solids into which the solid pentominoes will fit and, what is even more important, to discover other interesting shapes into which they can be assembled.

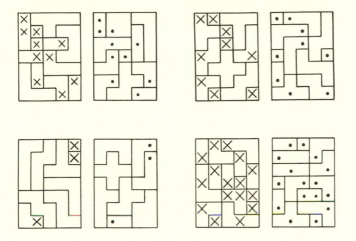

*Figure 115. Two
2 × 5 × 7 solids
built with solid
pentominoes.*

*Figure 116. The
two 2 × 5 × 7 solids
built with the planar
solid pentominoes
in one half.*

HIGHER DIMENSIONS AND RECTANGULAR TREES

It is not as difficult as one might imagine to list all the *n*-ominoes that can be formed in an *unlimited* number of dimensions, at least for small values of *n*; *n* connected cells never require more than *n* − 1 dimensions for their positioning, so that the pentominoes are the first case that go beyond three dimensions. Moreover, the construction of all *n*-ominoes requiring the full *n* − 1 dimensions is remarkably simple. When no limitation is imposed on the number of dimensions available, it is no longer reasonable to consider mirror twins as distinct; they can be rotated into coincidence in a space of higher dimension just as some of the one-sided pentominoes are distinct *in* the plane but can be flipped over into exact superposition if use is made of the third dimension.

For purposes of representation, it is convenient to replace each unit (a square, cube, or hypercube) by a single point, namely the point at its center, and to join these center points by lines if their corresponding cells are connected. Thus the tetromino of figure 117a is replaced by the "tree" of Figure 117b. In mathematical terminology, the tree is the *projective dual* of the polyomino.

*Figure 117. "Tree"
representation of a
tetromino.*

Chapter 6

To represent a large number of dimensions, it is further convenient to introduce a scheme in which connections in each dimension correspond to a different type of line (solid, dotted, dashed, and so on). Figure 118 shows all the distinct monominoes, dominoes, trominoes, tetrominoes, and pentominoes of unrestricted dimension, with the first dimension (horizontal) represented by the solid line, the second dimension (vertical) by a dashed line, the third dimension by a dotted line, and the fourth dimension by a line of alternating dashes and dots.

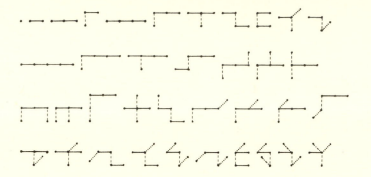

Figure 118. Monominoes through pentominoes of unrestricted dimension.

There are only seven tetrominoes in figure 118, rather than eight as in figure 112, because the last two tetrominoes of figure 112 are mirror images. Similarly, there are only twenty-three pentominoes of 1, 2, and 3 dimensions in figure 118 instead of the twenty-nine pieces identified by Klarner, because his pieces, used in figures 115 and 116, include six sets of mirror twins. The last three pentominoes in figure 118 are 4-dimensional.

There is an extensive mathematical literature on the enumeration of dimensional trees, which is effectively summarized in chapter 6 of John Riordan's book mentioned earlier. It is quite likely that the enumeration of "rectangular trees" of the type shown in figure 118 could be carried out by existing methods, where rectangular trees would be defined as *trees* (a set of points simply connected by lines) with the connecting lines variously "colored." The ordinary, or uncolored, trees are shown in figure 119, up through "order 6" (six points and five connecting lines).

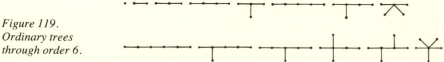

Figure 119. Ordinary trees through order 6.

In attempting to establish an equivalence between polyominoes and rectangular, colored trees, the following set of rules concerning the coloring must be observed:

1. At most, two lines of the same color may emanate from any single point.
2. Trees that differ only by a permutation (relabeling) of the colors are considered identical.
3. Every line segment must be assigned an "orientation" (that is, *plus* or *minus*), and trees of the same color but with different line-segment orientations are only equivalent if the orientations can be made to agree by reversing the sign of *all* segments having certain colors. (The orientations are conveniently represented by arrowheads on the line segments.)
4. In any linear sequence of line segments in a tree, it is not allowed to have the same number of plus-and-minus-oriented segments of each of the colors in the sequence. (The purpose of this rule is to prevent the tree from "closing back on itself" when interpreted as a polyomino.)

With this set of rules, the basic trees of figure 119 become the rectangular trees of figure 118, as shown for the monominoes, dominoes, trominoes, and tetrominoes in figure 120.

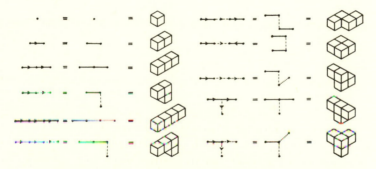

Figure 120.
Conversion of trees
to polyominoes.

Unfortunately, beyond this point the classification is no longer identical for polyominoes and for rectangular trees. The difficulty already encountered with certain pentominoes is illustrated by the four inequivalent trees in figure 121. As indicated, each of them "folds up" into the P pentomino.

Figure 121. Four
inequivalent trees
that fold up into the
same P pentomino.

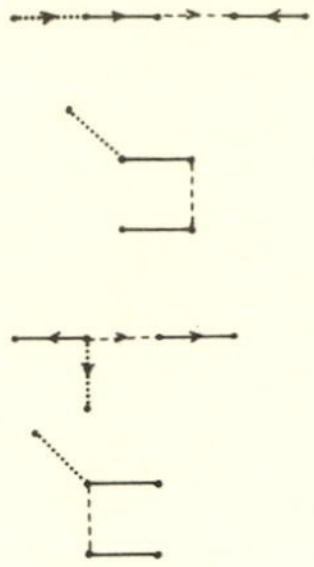

Figure 122.
Two different tree
representations
of the same solid
pentomino.

A similar illustration, involving the two distinct tree representations of the same solid pentomino, is shown in figure 122. Actually, these are the only cases of multiple-tree representation for the pentominoes. However, the situation gets progressively more complicated as the size of the *n*-ominoes increases. It may still be possible, despite these difficulties, to apply the enumeration techniques for trees to the problem of the number of solid *n*-ominoes that can be derived. The restriction to *plane* polyominoes merely restricts the trees to the bichromatic (two-colored) case.

One important case in which the colored trees are a completely faithful representation of the corresponding polyominoes is the study of the number of *n*-ominoes that require the maximum number, namely $n - 1$, of dimensions. In this case, each of the $n - 1$ edges of a tree with *n* points (or, as they are also known, *nodes*) must be colored a different color. This effectively eliminates all questions of symmetry and orientation. Thus, there is exactly one such *n*-omino for each ordinary or unrestricted tree with *n* nodes; and there are no other such *n*-ominoes. Hence, referring back to figure 119, there is one 1-dimensional domino, one 2-dimensional tromino, two 3-dimensional tetrominoes, three 4-dimensional pentominoes, and six 5-dimensional hexominoes. These are summarized in figure 123, where a heavy line has been introduced as the fifth color.

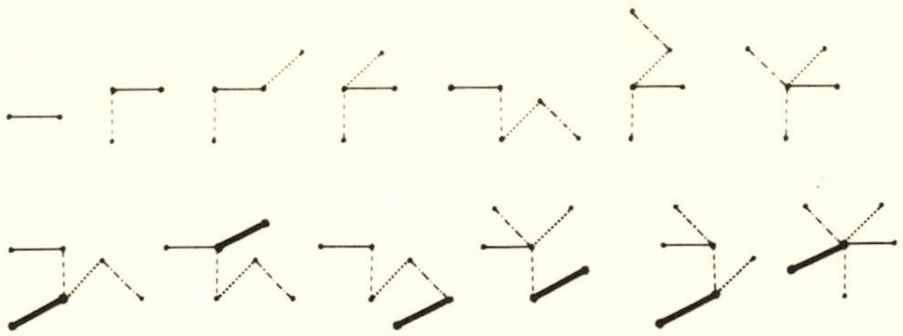

Figure 123. The
n-ominoes that
require n - 1
dimensions.

Generalizations of Polyominoes

ATTENTION will now be turned to *generalizations* of polyominoes, obtained first by relaxing the connectivity requirements and then by changing the basic building blocks from
squares to such shapes as triangles and hexagons. Finally, an
enumeration of the number of different ways in which a 2 ×
n rectangle can be filled with dominoes will be made.

PSEUDO- AND QUASI-POLYOMINOES

One way to generalize the notion of polyominoes is to relax
the requirements by which squares must be connected. As
mentioned in chapter 1, polyominoes may be regarded as being
"rookwise" connected. One can define pseudo-polyominoes
as kingwise connected sets of squares; that is, the king, which
can move diagonally as well as along rank and file, must be
able to reach any other square of the polyomino in a finite
number of moves. Even more general is the concept of quasi-
polyominoes, where a quasi-*n*-omino is any set of *n* squares
from a square planar array, irrespective of connectivity. These
concepts were first introduced by the author in "Checkerboards and Polyominoes" in the *American Mathematical
Monthly* in 1954.

The pseudo-*n*-ominoes for *n* = 1, 2, 3, and 4 are shown in
figure 124. The fact that the five pseudo-trominoes can be
fitted together into a 3 × 5 rectangle is illustrated in figure
125a. The only rectangles capable of including all twenty-two
of the pseudo-tetrominoes illustrated in figure 124 are the 8
× 11 and 4 × 22 rectangles. A simultaneous solution to both
problems is obtained by constructing two 4 × 11 rectangles,
as shown in figure 125b.

Obviously, pseudo-polyominoes can be extended to more
than two dimensions. Note that in three dimensions, there are
three solid pseudo-dominoes and fourteen solid pseudo-trominoes (not including mirror images) as shown in figure 126.
Unlike ordinary polyominoes, the number of pseudo-*n*-omi-
noes increases without limit as the dimension increases, even
for small values of *n*. The number of pseudo-dominoes in *D*
dimensions is exactly *D* (hence the two pseudo-dominoes in
figure 124 and the three in figure 126). The number of pseudo-

Chapter 7

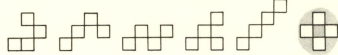

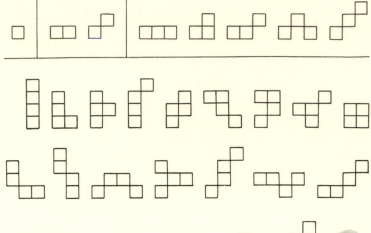

Figure 124. The
pseudo-n-ominoes
for n = 1, 2, 3,
and 4.

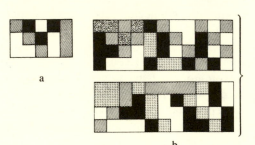

Figure 125. Fitting together
pseudo-n-ominoes to form rectangles:
(a) The 5 pseudo-trominoes can make
a 3 × 5 rectangle; (b) 2 congruent
rectangles constructed with the
22 pseudo-tetrominoes.

a

b

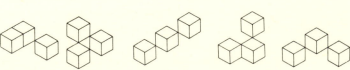

Figure 126.
Three dimensional
pseudo-polyominoes.

trominoes in D dimensions has not been studied, but the formula for them grows at least proportionally to D^4.

The method given earlier for obtaining all the ordinary $(n + 1)$-ominoes from the n-ominoes works equally well for pseudo-dominoes and, with any reasonable constraints on the dimensionality of the acceptable pieces, both for polyominoes and pseudo-polyominoes. In fact, precisely this method was used in arriving at figure 124 and figure 126. The method also applies to the trees of chapter 6, with and without colors and constraints.

The subject of quasi-polyominoes introduces some essentially new features. For one thing, it is no longer possible to draw all "simple" cases; for example, the number of quasi-dominoes in one dimension is already infinite. As a result, problems of *existence* and *constructability* are more typical than the more straightforward construction and enumeration problems encountered up to this point.

A construction problem involving quasi-trominoes was given in the article "Checkerboards and Polyominoes," referred to at the beginning of this chapter. It asked for a covering of the 8×8 checkerboard using twenty-one of the quasi-trominoes of figure 127a and one monomino. A solution is shown in figure 127c, making use of the hexomino indicated in figure 127b, which is assembled from two of the quasi-trominoes.

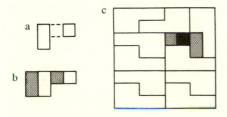

Figure 127. Covering a checkerboard with quasi-trominoes and a monomino.

Dr. Basil Gordon, professor of mathematics at the University of California, Los Angeles, investigated quasi-polyominoes in the following context: What is the quasi-polyomino of fewest cells, in D dimensions, that *cannot* be used to fill up all of D-dimensional space?

Gordon has solved this problem for $D = 1$, where the "space" is the infinite line with uniform segments represented in figure 128. The only quasi-monomino is the single segment (fig. 129a),

Figure 128. Uniform segments on the infinite line.

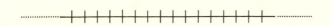

Chapter 7

and clearly this can be used repeatedly to fill up the "space" of figure 128. Every quasi-domino along the segmented line can be thought of in the form shown in figure 129b, where two unit segments are separated by an integral distance, t, for any $t = 0, 1, 2, 3, 4, \ldots ..$

It is clear that $t + 1$ of these quasi-dominoes can be juxtaposed to cover a solid stretch of length $2t + 2$, as indicated in figure 129c, where the same number (from 0 to t) identifies a component of the same quasi-domino. It is then a trivial matter to cover the entire "space" of figure 128 with an unlimited number of these solid stretches of length $2t + 2$.

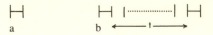

a   b

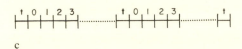

c

Figure 129. Covering the infinite line with quasi-dominoes.

Figure 130. The infinite line cannot be covered with this building block, without rotation.

For the case of quasi-trominoes, if the piece cannot be rotated, then the shape of figure 130 is an inadequate building block, since the gap it contains can never be filled. In the more interesting case where rotation *is* permitted, Gordon has proved the remarkable theorem that *every* quasi-tromino can be used to cover the 1-dimensional space of figure 128. For example, two quasi-trominoes of the type shown in figure 130 can be interlocked "back-to-back" to form a solid segment of length 6, which can be repeated to cover the infinite line. However, figure 131 shows a quasi-tetromino that clearly cannot succeed in covering the line. Thus, Gordon's result in one dimension is that every quasi-tromino, but *not* every quasi-tetromino, can be used to cover the infinite line.

Figure 131. A quasi-tetromino that cannot be used to cover the infinite line.

Considering two or more dimensions, Gordon has exhibited a quasi-polyomino of $3D$ cells (where D is the dimension) that cannot be used to cover, or "tile," the space. The idea is illustrated in figure 132, where the examples for $D = 2$ and $D = 3$ are shown. (Actually, this family of examples turns out

Figure 132. Quasi-polyominoes that cannot cover the infinite space.

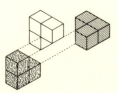

to be one of *pseudo*-polyominoes.) These examples are obtained by taking all 2-dimensional rook neighbors (rookwise adjacent) of an empty cell, plus D additional king neighbors (kingwise adjacent), which are rook neighbors to the original rook neighbors, two at a time. The empty central cell is thereby isolated. (Note that this construction does not work in one dimension.) Although it is thus proved that there exist quasi-polyominoes of $3D$ cells that will not tile D-dimensional space (for D greater than 1), it is certainly not proved that every quasi-polyomino of $3D - 1$ or fewer cells will tile the space. In fact, it is difficult to show more than that all quasi-polyominoes of $D + 1$ cells will tile D-dimensional space.

The mathematical logician Dr. E. F. Moore of the Bell Telephone Laboratories once considered a variety of problems involving the covering or tiling of D-dimensional space using an unlimited number of replicas of a finite set of "standard parts." He has formulated a challenging and profound conjecture concerning all such problems, including the quasi-polyomino covering problems just discussed. Moore's conjecture was that whenever a finite set of standard parts has the property that an unlimited number of replicas of these parts can be used to cover D-dimensional space, there is some *finite pattern* that they can cover, which can then be repeated periodically to finish covering D-dimensional space. A supporting instance of this conjecture was given in figure 129, where the arbitrary quasi-domino was first used to fill up a solid segment, and this solid segment was then repeated periodically to cover the infinite line. Moore's conjecture did not assert that every infinite covering (one that covers the entire plane) is in fact a periodic covering (a covering of the entire plane by periodic repetition of a basic pattern). Such an assertion would be quite false. The conjecture does assert that if an infinite covering exists, then a periodic covering also exists. This conjecture appears to be exceedingly difficult to prove or disprove, and Moore even advanced the suggestion that it may be logically undecidable. However, a counter-example was found by Moore and Hao Wang, professor of logic, Harvard University, demonstrating that certain sets of shapes can fill the plane, although they cannot fill it periodically.

The reader is invited to test the twelve ordinary, familiar pentominoes to see which of them can be used, repeating only the one shape chosen, to tile the plane. A distinction can be made between those that tile *without* allowing rotation, and those that can tile the plane only if rotated positions are allowed. An example of the latter case is shown in figure 133.

Chapter 7

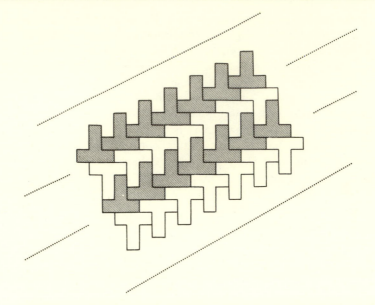

Figure 133. Tiling the plane with the T pentomino.

TRIANGULAR AND HEXAGONAL ANIMALS

In his writings on polyominoes, Dr. Frank Harary, formerly professor of mathematics at the University of Michigan, has referred to *n*-ominoes as "*n*-celled animals." In his *American Mathematical Monthly* article, "Checkerboards and Polyominoes," the author suggested that other regular tilings (either triangular or hexagonal) could be used for the board and for the objects covering it. Connected combinations of equilateral triangles can be called *triangular animals* and, similarly, connected hexagons called *hexagonal animals*. The triangular animals of size *n* for *n* = 1, 2, 3, 4, and 5 are shown in figure 134.

Figure 134. Triangular animals of size n up through n = 5.

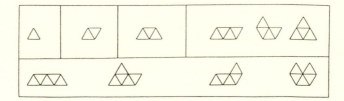

In an article, "Maestro Puzzles" (a name under which the pentomino puzzle is marketed commercially in England), in the *Mathematical Gazette*, the English mathematicians J. E. Reeve and J. A. Tyrrell considered patterns that can be formed using all twelve triangular animals six triangles in size. They report having discovered more than forty solutions to the covering of the rhombus shown in figure 135, which exhibits one of these solutions.

*Figure 135. A rhombus covered with the
12 triangular animals of size 6.*

Reeve and Tyrrell also consider the problem of arranging nine of these twelve triangular animals into the regular hexagon of figure 136. They point out that there are $\binom{12}{9} = 220$ ways to decide which nine of the twelve to use, and that ninety of these 220 possible solutions are clearly impossible because they involve using one but not both of the two "unbalanced" animals of figure 137. Since the entire hexagon of figure 136 is "balanced" in alternating colors, it is clear that it cannot be covered if an odd number of the unbalanced animals are used.

*Figure 136. A
regular hexagon.*

| Balanced Animals | Unbalanced Animals |

*Figure 137.
The balanced and
unbalanced
hexagonal animals.*

Reeve and Tyrrell assert in their article that twenty-two of the remaining 130 problems are insoluble, "but [they] have been unable to find a simple proof of this." The reader is encouraged to try his hand at these problems, and to investigate other interesting patterns into which triangular animals may or may not fit. For example, can the seven animals of four and five triangles be fitted into the rhombus shown in figure 138?

In his column "Puzzles and Paradoxes" in the British periodical *New Scientist*, T. H. O'Beirne has devoted considerable space to problems involving the triangular animals. Calling two equilateral triangles joined to form a rhombus a *diamond*, he generalizes this to triamonds, tetriamonds, pentiamonds, hexiamonds, and so on. He has named the twelve *hexiamonds* (six connected triangles) as shown in figure 139.

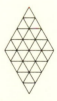

*Figure 138. A
rhombus.*

Chapter 7

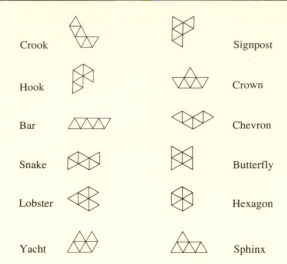

Crook

Signpost

Hook

Crown

Bar

Chevron

Snake

Butterfly

Lobster

Hexagon

*Figure 139. The
12 hexiamonds.*

Yacht

Sphinx

*Figure 140. A
honeycomb to be
filled with the
19 one-sided
hexiamonds.*

Many of O'Beirne's constructions involve the use of nineteen hexiamonds—the mirror images of the seven asymmetric hexiamonds are added to the twelve shapes already considered. One problem is to fit these nineteen pieces into three hexagons, one being the hexagonal piece itself, and each of the other two being the hexagon shown in figure 136. Also discussed at length are problems of filling up the honeycomb shape of figure 140.

There are twenty-four *heptiamonds* (seven connected triangles), as shown in figure 141. The reader is encouraged to look for interesting configurations that this set of shapes will cover.

*Figure 141. The
24 heptiamonds.*

If a dot is placed at the center of each cell in the square tiling of the plane, the vertices of a similar square tiling appear. However, if this is done to triangular tiling, the vertices of hexagonal tiling appear, and the dots in the center of hexagonal tiling indicate vertices of triangular tiling. The triangular and hexagonal tilings are therefore dual configurations, whereas the square tiling is self-dual. As shown in figure 136, the triangular tiling admits a checkerboard coloring. This is not possible with hexagonal coverage. However, there is a three-color analog of the checkerboard coloring for hexagonal tiling, as illustrated in figure 142.

The simplest hexagonal animals have been studied by chemists—although not under that name—as models of the molecular structure of organic compounds. Hexagonal animals of 1, 2, 3, 4, and 5 hexagons are shown in figure 143. There had been no previously published problem literature on this subject in 1965, so the reader was invited to propose configurations for these animals. Now, however, see pp. 122–126 for *polyhex* problems.

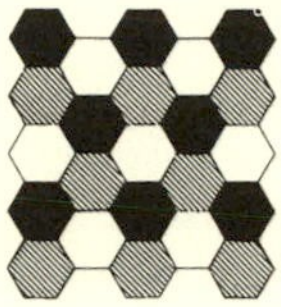

Figure 142.
Three-color
hexagonal tiling.

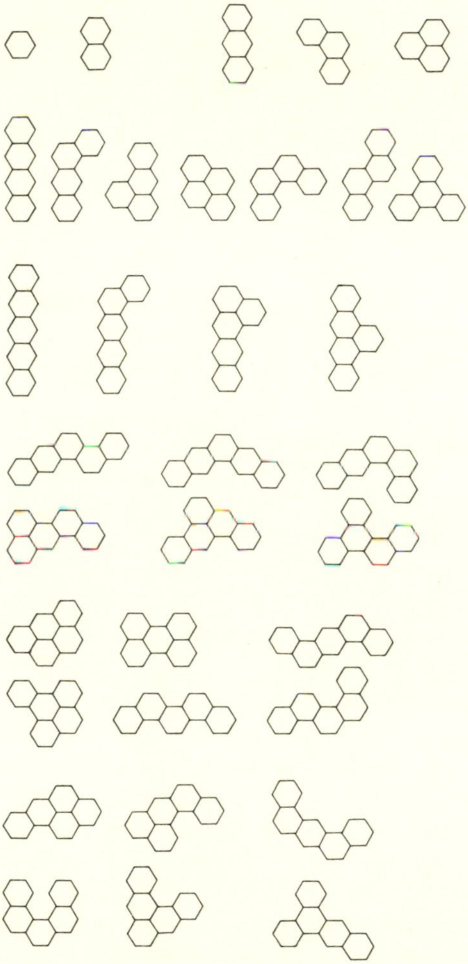

Figure 143.
The hexagonal
animals up through
size 5.

RECTANGLES OF DOMINOES

Another type of polyomino problem is to derive mathematical expressions for the number of ways that a given set of polyominoes can fill up a specified region. Obviously, this is extremely difficult to do in general, but a special case in which the answer is rather easy to obtain and is quite elegant will now be presented.

In the Problem Section of the May 1961 issue of *American Mathematical Monthly*, W. E. Patton proposed the following problem: "It is desired to form a $2 \times n$ rectangle from 1×2 rectangles (dominoes), or we may say, to cover the rectangle with dominoes. In how many distinct ways can this be done, where two solutions are distinct when they cannot be brought into coincidence by rotations and reflections?"

This author's solution to Patton's problem (published in the magazine's January 1962 issue) goes as follows: First, all $2 \times n$ rectangles of dominoes are enumerated without regard to possible symmetries among them. For $n = 1$, the number of possible rectangles is $f_1 = 1$, and for $n = 2$, the two dominoes are either horizontal or vertical, so that $f_2 = 2$. For n greater than 2, starting at the left, the $2 \times n$ rectangle begins with either a vertical domino, which can be "extended" to fill out the $2 \times n$ rectangle in f_{n-1} ways, or with a pair of horizontal dominoes, which can be "extended" to fill out the rectangle in f_{n-2} ways. Hence $f_n = f_{n-1} + f_{n-2}$, as illustrated in figure 144.

Figure 144. Filling a $2 \times n$ rectangle with dominoes for $n = 1, 2, 3,$ and 4.

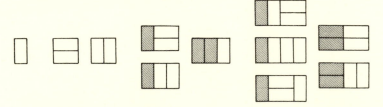

The sequence beginning $f_1 = 1$, $f_2 = 2$ and satisfying $f_n = f_{n-1} + f_{n-2}$ for n greater than 2 is the famous *Fibonacci sequence*: 1, 2, 3, 5, 8, 13, 21, 34, 55, 89, ... The thirteenth-century Italian mathematician Leonardo Fibonacci (also known as Leonardo of Pisa) first introduced this sequence to the mathematical world. He started with 0 and 1, then added these two terms together to get 1 again, and continued to generate each new term as the sum of its two immediate predecessors. Thus, in its original form, the sequence began: 0, 1, 1, 2, 3, 5, 8, 13, ...

Let C_m refer to the number of solutions to the $2 \times m$ rectangle in which left-to-right mirror images are *not* regarded as distinct. Then $C_m = \frac{1}{2}(f_m + s_m)$, where s_m is the number of solutions that are left-to-right symmetric. This result is a simple application of the formula $N = \frac{1}{2}(T + C)$ of chapter 5, used to count the number of distinct cases under an "involution." (As a direct derivation, the number that are not left-to-right symmetric is clearly $f_m - s_m$, and only half of these asymmetric ones are distinct when left-to-right mirror images are considered to be the same. However, all the symmetric cases retain their individual identity, for a total of $\frac{1}{2}(f_m - s_m) + s_m = \frac{1}{2}(f_m + s_m)$.)

To evaluate s_m, consider odd and even m separately. For $m = 2n + 1$, a left-to-right symmetric solution must have a vertical domino in the center, leaving a $2 \times n$ rectangle on each side. One side can be completed in f_n ways, and then the other side is completely specified as the mirror image. Thus $s_{2n+1} = f_n$.

For $m = 2n$, a vertical line down the center of the rectangle either cuts a horizontal pair of dominoes or cuts no dominoes. In the former case there are f_{n-1} ways to specify the $2 \times (n - 1)$ rectangle to the left of the horizontal pair; in the latter case, f_n ways to specify the $2 \times n$ rectangle to the left of the midline; and, having specified the left half, one can determine the right half by mirror symmetry. Hence, $s_{2n} = f_{n-1} + f_n = f_{n+1}$.

Thus, $C_{2n+1} = \frac{1}{2}(f_{2n+1} + f_n)$ and $C_{2n} = \frac{1}{2}(f_{2n} + f_{n+1})$, which is the answer to the above problem except for the case $m = 2$, since the 2×2 rectangle is a square and admits of a 90-degree rotational symmetry, which reduces $C_2 = 2$ to $C'_2 = 1$, without affecting any other cases.

Table 5 summarizes the result.

TABLE 5

n	Number of Covered $2 \times n$ Rectangles
1	1
2	1
3	$\frac{1}{2}(f_3 + f_1) = \frac{1}{2}(3 + 1) = 2$
4	$\frac{1}{2}(f_4 + f_3) = \frac{1}{2}(5 + 3) = 4$
5	$\frac{1}{2}(f_5 + f_2) = \frac{1}{2}(8 + 2) = 5$
6	$\frac{1}{2}(f_6 + f_4) = \frac{1}{2}(13 + 5) = 9$
7	$\frac{1}{2}(f_7 + f_3) = \frac{1}{2}(21 + 3) = 12$
8	$\frac{1}{2}(f_8 + f_5) = \frac{1}{2}(34 + 8) = 21$
9	$\frac{1}{2}(f_9 + f_4) = \frac{1}{2}(55 + 5) = 30$
10	$\frac{1}{2}(f_{10} + f_6) = \frac{1}{2}(89 + 13) = 51$
11	$\frac{1}{2}(f_{11} + f_5) = \frac{1}{2}(144 + 8) = 76$
12	$\frac{1}{2}(f_{12} + f_7) = \frac{1}{2}(233 + 21) = 127$

The number of coverings of the $3 \times n$ rectangle with dominoes is clearly a much more difficult problem, but the reader is invited to try. A much easier problem is to enumerate the number of ways to cover the $2 \times n$ rectangle, or even the $3 \times n$ rectangle, with right trominoes; the number of ways to cover the $4 \times n$ rectangle with right trominoes appears to be a challenging problem with reasonable hope of an attainable solution. The reader is encouraged to attempt it.

A solution to the number of coverings of the $3 \times n$ rectangle with dominoes is given in chapter 7 of the book *Concrete Mathematics* by Graham, Knuth, and Patashnik, which also contains references to the extensive mathematical physics literature on the "dimer problem," an important application of polyomino theory.

Tiling Rectangles with Polyominoes

IN 1968, David A. Klarner defined the *order n* of a polyomino *P* as the minimum number of congruent copies of *P* that can be assembled (allowing translation, rotation, and reflection) to form a rectangle. For those polyominoes which will not tile any rectangle, the order is undefined. A polyomino has *order 1* if and only if it is itself a rectangle.

A polyomino has *order 2* if and only if it is "half a rectangle," since two identical copies of it must form a rectangle. In practice, this means that the two copies will be 180-degree rotations of each other when forming a rectangle. Some examples are shown in figure 145.

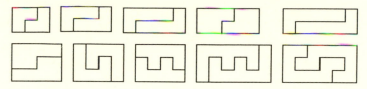

Figure 145. Some polyominoes of order 2.

There are no polyominoes of order 3. (This result has been proved by Ian Stewart of the University of Warwick, in England, in a 1992 paper with A. Wormstein.) In fact, the only way any rectangle can be divided up into three identical copies of a "well-behaved" geometric figure is to partition it into three *rectangles* (see fig. 146), and by definition a rectangle has order 1.

Figure 146. How three identical rectangles can form a rectangle.

There are various ways in which four identical polyominoes can be combined to form a rectangle. One way, illustrated in figure 147, is to have four 90-degree rotations of a single shape forming a square.

Figure 147. Polyominoes of order 4 under 90-degree rotation.

Another way to combine four identical shapes to form a rectangle has the four-fold symmetry of the rectangle itself: left-right, up-down, and 180-degree-rotational symmetry. Some examples of this appear in figure 148.

Figure 148.
Polyominoes of
order 4 under
rectangular
symmetry.

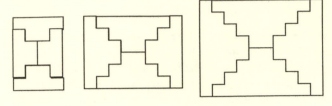

More complicated order-4 patterns were also found by Klarner, some of which are illustrated in figure 149.

Figure 149.
Another order-4
construction
by Klarner.

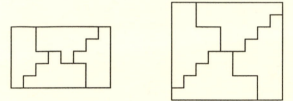

Beyond order 4, there is a systematic construction (found by Golomb in 1987) which gives examples of order 4s for every positive integer s; and eight isolated examples of small polyominoes with respective orders 10, 18, 24, 28, 50, 76, 92, and 312 are known.

Figure 150 shows the isolated examples of order 10 (Golomb 1966), and orders 18, 24, and 28 (Klarner 1969).

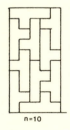

Figure 150. Four "sporadic" polyominoes, of orders 10, 18, 24, and 28, respectively.

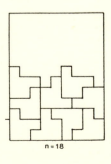

n=10

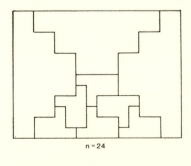

n = 18

n = 24

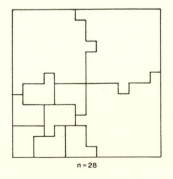

n = 28

Figure 151 shows the example of order 50, found by William Rex Marshall of Dunedin, New Zealand, in 1990.*

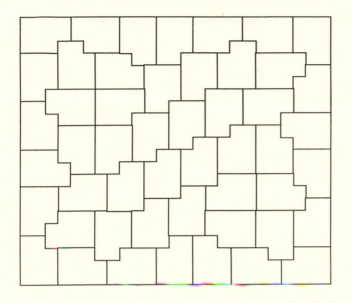

Figure 151. An 11-omino of order 50.

Figure 152 shows the examples of orders 76 and 92, both found by Karl A. Dahlke in 1988, but first discovered by T. W. Marlow in 1985.

Figure 152. A heptomino of order 76 and a hexomino of order 92.

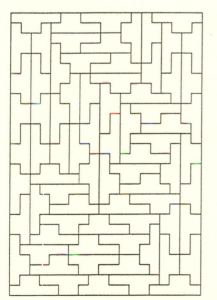

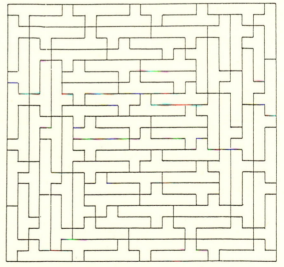

*William Rex Marshall, private communications dated 14 May 1990 and 25 November 1991.

The heptomino of order 76 in figure 152 cannot tile its minimum rectangle symmetrically. This is also true of the dekomino in figure 153 of order 96, whose minimum rectangle (the 30 × 32) was discovered by William Rex Marshall in 1991.

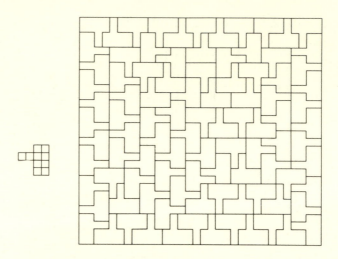

Figure 153. A dekomino of order 96.

Finally, figure 154 shows the example of order 312 (Dahlke 1989), although in this case it is not absolutely certain that no smaller number of copies of the octomino in question will form a rectangle.

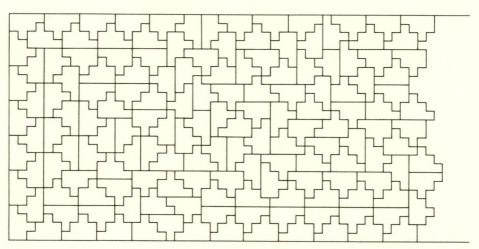

Figure 154. An example of order 312.

No polyomino whose order is an odd number greater than 1 has ever been found, but the possibility that such polyominoes exist (with orders greater than 3) has not been ruled out.

The known *even* orders of polyominoes are all the multiples of 4, as well as the numbers 2, 10, 18, and 50. Curiously, these even orders which are not multiples of 4 all exceed multiples of 8 by two. Whether there are other even orders, and what they might be, is still unknown. The smallest even order for which no example is known is order 6. Figure 155 shows one way in which six copies of a polyomino can be fitted together to form a rectangle, but the polyomino in question (as shown) actually has order 2.

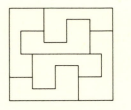

Figure 155. A 12-omino of order 2 which suggests an order-6 tiling.

The Golomb construction for polyominoes of order 4s gives its first new example, order 8, when $s = 2$. The underlying tiling concept of how to fit eight congruent shapes together to form a rectangle is shown in figure 156.

Although the shape used in figure 156 is not a polyomino, the same concept can be realized using the 12-omino shown in figure 157.

Figure 156. A rectangle formed from eight congruent pieces.

Figure 157. A polyomino of order 8.

To show that there are infinitely many dissimilar polyominoes having order 8, we can form the family of polyominoes shown in figure 158. For each integer $r \geq 1$, this construction produces a $(3r^2 + 6r + 3)$-omino of order 8, and clearly no two of these are similar.

It is also easy to show that none of these polyominoes can have order less than 8. The proof begins by observing that only the "heel of the boot" can be in any *corner* of the rectangle to be tiled. Then the "toe" of the boot must be mated with the notch at the top-back of another boot. The quickest way to finish off the rectangle then requires eight copies of the polyomino.

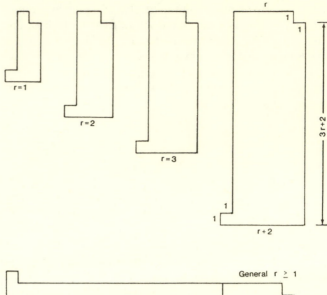

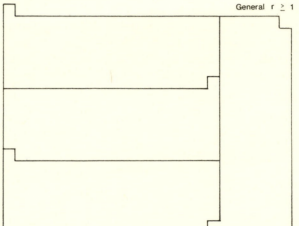

*Figure 158. Dis-
similar polyominoes
of order 8, and how
to stack them.*

In figure 159, we see a construction for a polyomino of order $n = 4s$ for every $s = 1, 2, 3, 4, \ldots$.

(Starting with a 2 by $4s - 2$ rectangle, we remove a single square from one corner and attach it as a "toe" at the opposite corner, to obtain the polyomino of order $4s$.)

The idea shown in figure 158 can be applied not only to order $n = 8$, but to any order $n = 4s$, to obtain infinitely many dissimilar polyominoes of order $4s$. The general construction involving both r and s begins with a rectangle which is $(r + 1) \times (2s - 1)(r + 1)$, and moves a single 1×1 square from the top-back of the "boot" to become a "toe" at the opposite

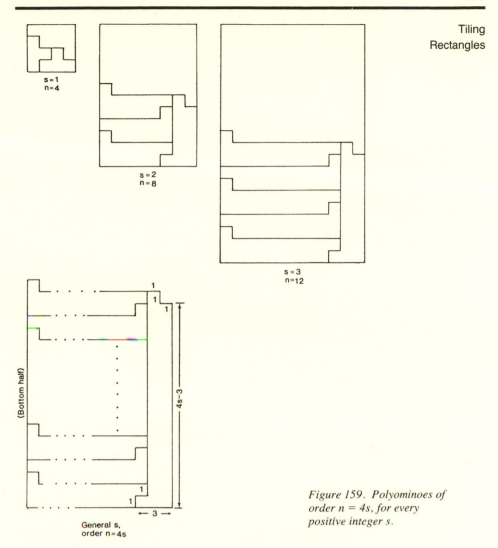

s=1
n=4

s=2
n=8

s=3
n=12

(Bottom half)

4s-3

1
1
1

1
1

← 3 →

General s,
order n=4s

*Figure 159. Polyominoes of
order n = 4s, for every
positive integer s.*

corner. (The proof that the resulting figure truly has order n = $4s$ is analogous to the proof given for $n = 8$.)

Actually, there are many different shapes that can be removed from the top-back of the "boot," and then affixed to form the "toe," some of which are illustrated in figure 160 for the case $s = 2$, $r = 3$. The necessary and sufficient condition for the "toe" to work is that it be symmetric in the diagonal from upper left to lower right, and that its removal from the top right of the rectangle does not disconnect the figure.

Chapter 8

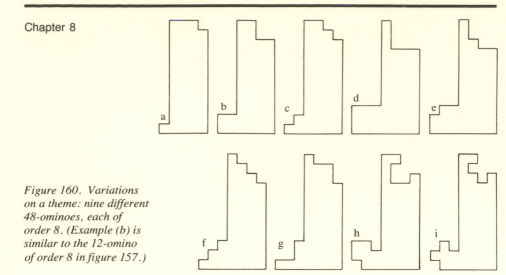

Figure 160. Variations on a theme: nine different 48-ominoes, each of order 8. (Example (b) is similar to the 12-omino of order 8 in figure 157.)

Klarner called a polyomino P *odd* if it is *possible* to use an odd number of copies of P to form a rectangle (not necessarily the *minimum* rectangle for P). He showed that

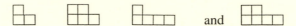 and

are all odd, being able to form rectangles made of 15, 21, 27, and 11 copies, respectively, as shown in figure 161.

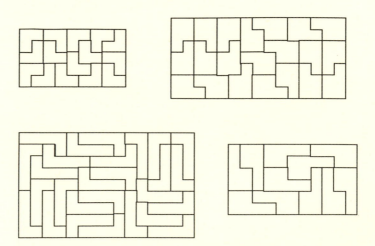

Figure 161. Four "odd" polyominoes.

He also showed that fifteen copies of any polyomino which is three quadrants of a rectangle ( ) can be used to pack a rectangle, using the tiling in figure 161 for the

tromino 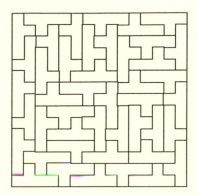 . All of these odd polyominoes have order 2.

Klarner asked whether the order-10 pentomino 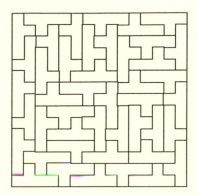 is

odd. This was answered in the affirmative in 1974 by Dr.
Jennifer Haselgrove of Glasgow, Scotland, as shown in figure
162, where forty-five copies of the pentomino are used to form
a 15 × 15 square.

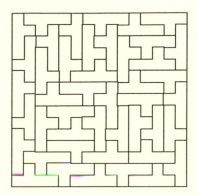

*Figure 162. A pentomino
of order 10 has
"odd-order" 45.*

In 1991, William Rex Marshall discovered that 141 copies
of his 11-omino of order 50, shown in figure 151, can be used
to tile a 33 × 47 rectangle. This is shown in figure 163.

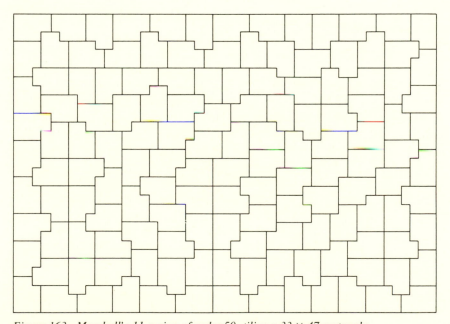

Figure 163. Marshall's 11-omino of order 50, tiling a 33 × 47 rectangle.

If we call the smallest odd number of copies of the odd polyomino P that will form a rectangle the *odd-order* of P, then the odd-orders that are known to occur are 1, 11, 15, 21, 27, 45, and 141. Does any nonrectangle odd polyomino have odd-order less than 11? Does any nonrectangle odd polyomino have an odd-order equal to its order? Besides Klarner's construction, another construction that yields infinitely many odd polyominoes was recently found by Michael Reid, a graduate student at the University of California, Berkeley. Reid's family is illustrated in figure 164.

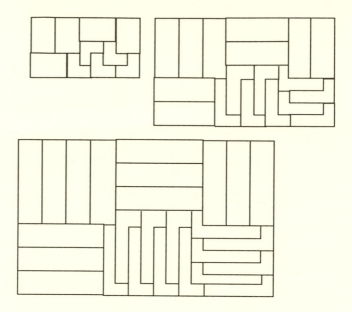

Figure 164.
Michael Reid's
infinite family
of "odd"
polyominoes.

What odd numbers can occur as odd-orders of polyominoes? (Because no polyomino has order 3, no odd polyomino can have odd-order 3.) These are all open questions.

If a polyomino has no order (i.e., if it cannot tile any rectangle), it may still be able to tile the entire plane, or various subregions of the plane, such as an infinite strip or a bent strip. Such tilings are illustrated in figure 165, using the X pentomino, the F pentomino, and the N pentomino, respectively.

In figure 166, we see a tiling hierarchy for polyominoes (Golomb, 1966). A polyomino that can tile any of the regions specified in the hierarchy can also tile all the regions lower in the hierarchy. Thus, the "true category" of a polyomino is the *highest* box in the hierarchy that it can occupy. Most of this chapter has been concerned with polyominoes that occupy the highest box—i.e., they can tile rectangles. The "true categories" known to have members are: Rectangle, Bent Strip,

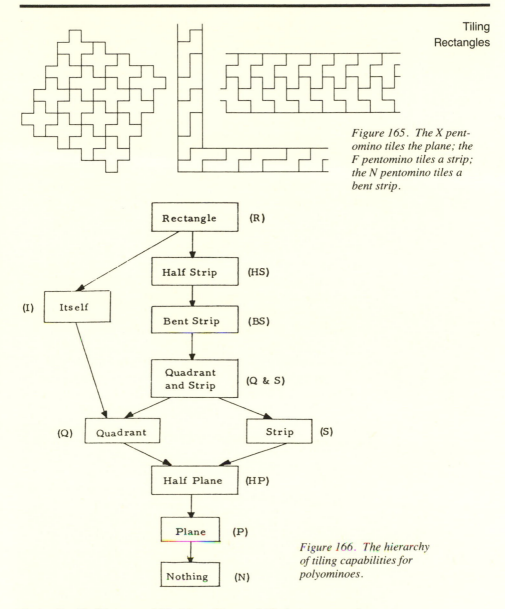

Figure 165. The X pent-
omino tiles the plane; the
F pentomino tiles a strip;
the N pentomino tiles a
bent strip.

Figure 166. The hierarchy
of tiling capabilities for
polyominoes.

Strip, Itself, Plane, and Nothing. Figure 167 shows examples
of the categories "Itself" and "Nothing." (The others have
already been illustrated.) For each of the other positions in the

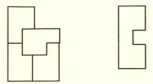

Figure 167. A 5-omino which
tiles "itself," and a 9-omino
which cannot tile the plane.

Chapter 8

hierarchy, it is an open question whether any polyomino has that position as its "true category."

In 1962, I wrote several articles about figures that can be carved into congruent copies of themselves, like the P pentomino in figure 167. Since such figures can *tile* the plane and are composed of *replicas* of themselves, I coined the term *reptiles* for this class of geometric figures. All this was popularized by Martin Gardner in a 1963 "Mathematical Games" column in *Scientific American,* and one of my articles ("Replicating Figures in the Plane") appeared in 1964 in *Mathematical Gazette*. However, since most polyominoes are not rep-tiles, and most rep-tiles are not polyominoes, I will leave the development of this fascinating subject for another book.

If we allow tilings using a *finite set* of polyomino shapes as the tiles, we can provide examples for each of the remaining positions in the hierarchy. Specifically, we can exhibit sets of only *two* polyominoes such that the "true category" for that set is either the Half-Plane, the Quadrant, the Quadrant and Strip, or the Half-Strip. Examples of such sets of two polyominoes, and the corresponding tilings, are shown in figures 168, 169, 170, and 171, respectively.

To determine whether a given arbitrary polyomino can tile any rectangle or whether a given set of polyominoes can be used to tile the plane (or similar problems of determining the "true category" of one or a set of polyominoes) is known, in general, to be "computationally undecidable." This means that there is no computer program that, given the arbitrary poly-

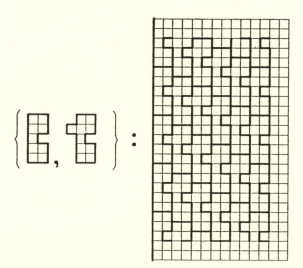

Figure 168. A characteristic example for the Half-Plane.

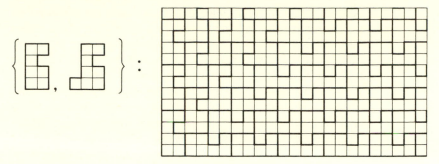

Figure 169. A characteristic example for the Quadrant.

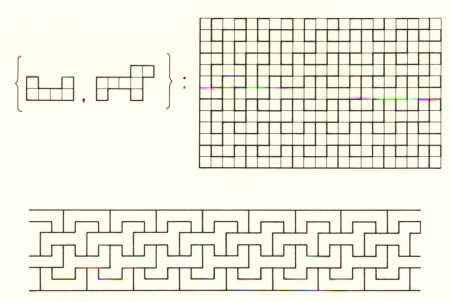

Figure 170. The dog-and-trough characteristic example for the Quadrant-and-Strip.

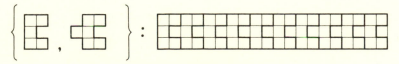

Figure 171. A characteristic example for the Half-Strip.

omino (or set of polyominoes) as input, can be guaranteed to give a yes-or-no answer to "does it tile any rectangle?" (or, "does it have a specified position in the tiling hierarchy as its true category?") in a specified length of time, where the specified length of time is allowed to depend, in any deterministic way, on the given polyomino (or, set of polyominoes). One of the consequences of this very deep result is that there is no "computable function of n" which provides an upper bound to the possible order of every n-omino. (This follows from the fact that if we knew that every n-omino that can tile any rectangle has an order not exceeding n^{n^n}, for example, it would be a finite search, in a predictable—albeit enormous—amount of time, to try all possible arrangements of no more than n^{n^n} copies of the given polyomino to see if any rectangle is possible.) We have seen that there is a tetromino of order 4, a pentomino of order 10, a hexomino of order 92, and an octomino believed to have order 312. However, there is no expression involving n that one can write down (such as n^{n^n}) which will guarantee that every n-omino that tiles a rectangle will have an order not exceeding the value of that expression, evaluated at n, for all values of n.

Some Truly Remarkable Results

GOMORY'S THEOREM

We started this book by proving that if two opposite corner squares are removed from an 8 × 8 checkerboard, what is left cannot be covered with dominoes. The same proof shows that if any two squares *of the same color* are removed from an 8 × 8 checkerboard, what remains cannot be covered with dominoes.

The converse of this theorem was proved by Ralph Gomory, president of the Alfred P. Sloan Foundation and formerly vice president for research at IBM: If any two squares of *opposite* color are removed from an 8 × 8 checkerboard, what is left can *always* be covered with dominoes. Here is his ingenious proof: Impose the barriers shown in figure 172 between squares of the checkerboard. No domino is allowed to cross a barrier.

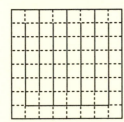

Figure 172. Gomory's barriers on the 8 × 8 checkerboard.

These barriers force any domino covering to follow a "single file" circuit on the board. If two opposite-colored squares are removed, the 64-square circuit is cut into two *even-length* pieces (or only one even-length piece if the two deleted squares are consecutive in the circuit); and any even-length path within the circuit can be covered (in exactly one way!) with dominoes. Three examples are shown in figure 173.

Figure 173. Examples of domino coverings following Gomory's barriers when two opposite-colored squares are removed from the checkerboard.

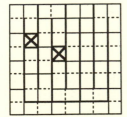
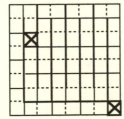
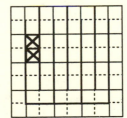

Chapter 9

Gomory's barrier pattern in figure 172 is not the only one that can be used in this proof. There are literally hundreds of others, four of which are shown in figure 174.

Figure 174. A few of the other barrier patterns that can be used in proving Gomory's theorem.

PATTERNS OF OCTOMINOES AND BEYOND

There are 369 octominoes, six of which have holes. In the period of 1965 to 1967, David Bird of North Shields, England, discovered a number of incredible patterns, each using all 369 octominoes once each, with the holes aligned symmetrically! Without *extra* holes, these patterns each have an area of 369 × 8 + 6 = 2958, which factors as 2 × 3 × 17 × 29. Accordingly, Bird formed rectangles of sizes 51 × 58 (fig. 175), 34 × 87 (fig. 176), 29 × 102 (not shown here), 17 × 174 (not shown here), and *three* 29 × 34 rectangles (fig. 177). He also formed a jagged square with thirteen holes arranged symmetrically (fig. 178), and a 55 × 55 highly symmetric square pattern (fig. 179). To keep these constructions from being too easy, he usually required his patterns to have no "crossroads," that is, no instance of four octominoes having a common vertex. (Of the five octomino patterns reproduced here, only figure 178 allows "crossroads.")

In 1972–1973, David Bird composed several patterns using all 1285 enneominoes (another word for "nonominoes") of which 36 have a single hole, and one has a double (domino-shaped) hole. Two of his patterns embedded these shapes in a 109 × 109 square, with all the holes arranged symmetrically! One of these, which he dated 6 March 1973, is shown

Figure 175. The 369 octominoes in a 51 × 58 rectangle.

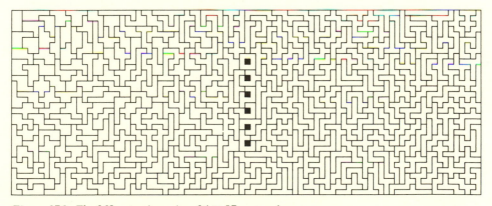

Figure 176. The 369 octominoes in a 34 × 87 rectangle.

Chapter 9

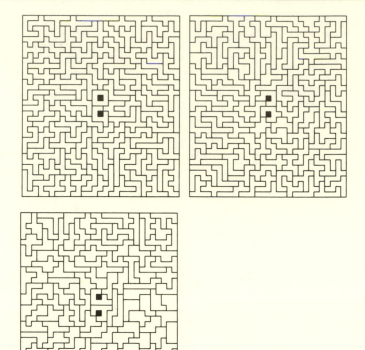

Figure 177. The 369 octominoes in three congruent rectangles, each 29 × 34.

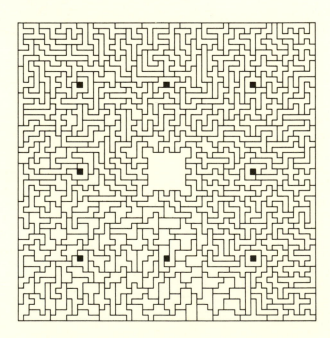

Figure 179. The 369 octominoes in a pattern of maximum symmetry. Solid internal corners. No crossroads. No segment longer than the straight piece.

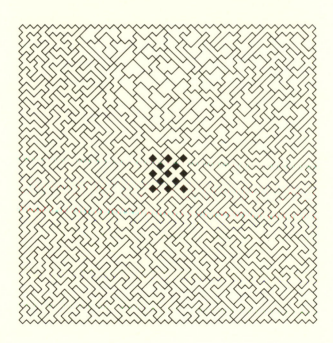

Figure 178. Two symmetrical patterns made with the 369 octominoes.

in figure 180. In his cover letter of 10 March 1973, he wrote: "I am, so far as I know, the only person to have composed patterns with the 369 octominoes, the 1285 enneominoes, the 66 octiamonds and the 160 enneiamonds."*

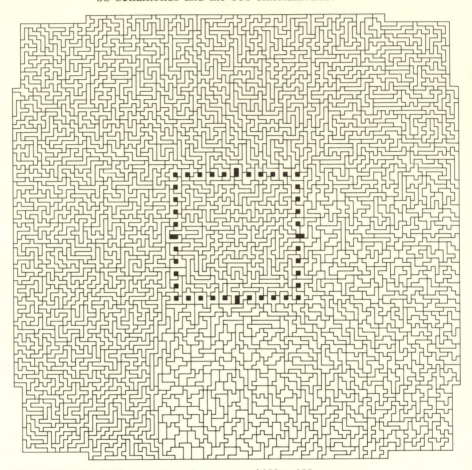

Figure 180. The 1285 enneominoes in a truncated 109 × 109 square.

DE BRUIJN'S THEOREM

In a 1969 article, Professor Nicolas G. de Bruijn of the Eindhoven Technical University in the Netherlands proved a remarkable result: an $a \times b$ rectangle cannot be tiled using 1 $\times$ n tiles unless n divides one of the sides of the rectangle. Thus, an 8 × 9 floor cannot be tiled with 1 × 6 tiles, even though the area of the tile (6) divides the area of the floor to

*David Bird, private communications dated from 28 October 1966 to 10 March 1973.

be tiled (72), since 6 divides neither 8 nor 9. He also proved the corresponding result in three dimensions (a rectangular box of sides a, b, and c cannot be completely filled with $1 \times 1 \times n$ blocks unless n divides one of the sides of the box), and in all higher dimensions as well. His ingenious proof makes use of a property of the complex numbers, which led others to look for more "elementary" proofs. In 1987, Professor Stan Wagon, then at Smith College, published an article with fourteen different proofs of de Bruijn's theorem, but did not include my proof by a coloring argument, which goes as follows.

We have an $a \times b$ board, and would like to cover it with $1 \times n$ tiles. Color the board in n colors, applying the colors consecutively and periodically on the diagonals of the board. (This is illustrated in figure 181 for the case $a = 8$, $b = 9$, $n = 6$.) Since a $1 \times n$ tile will cover each of the n colors exactly once, wherever it is placed, the entire board cannot be covered exactly with $1 \times n$ tiles unless all n colors occur equally often on the $a \times b$ board.

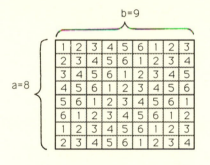

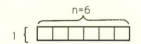

Figure 181. The diagonal coloring, illustrated for $a = 8$, $b = 9$, $n = 6$.

If neither a nor b is divisible by n, then divide a by n to get a remainder a', with $0 < a' < n$, and divide b by n to get a remainder b' with $0 < b' < n$. Except for an $a' \times b'$ rectangle in the corner of the original $a \times b$ board, the *rest* of the big board can easily be covered with $1 \times n$ tiles (as illustrated in figure 182, again with $a = 8$, $b = 9$, $n = 6$).

Figure 182. The diagonal coloring, and the reduced rectangle, illustrating the proof of de Bruijn's theorem for the case $a = 8$, $b = 9$, $n = 6$.

Since the *n* colors appear equally often under each of the tiles, they must also occur equally often in the reduced $a' \times b'$ rectangular board, in order for the original board to have each of the *n* colors equally often. Suppose a' is less than or equal to b'. Then in the reduced board there will be at least one solid-color diagonal of length a'. Since there is at least one color occurring at least a' times, and we require all *n* colors to occur equally often on the reduced ($a' \times b'$) board, the reduced board must have at least $a' \times n$ squares; but this is *larger* than the area ($a' \times b'$) of the reduced board, since b' is less than *n*. (The identical reasoning also works if b' is less than or equal to a'.) Hence the covering with $1 \times n$ tiles is impossible if neither *a* nor *b* is a multiple of *n*.

Besides the reduced 2×3 board in figure 182, the example of a reduced 8×5 board (for $a = 28$, $b = 35$, $n = 10$) is shown in figure 183, to help the reader verify the reasoning in the proof just given.

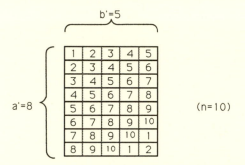

Figure 183. The reduced rectangle for the case a = 28, b = 35, n = 10. Here a' = 8 and b' = 5, with b' being the smaller side. There are colors (colors 5, 6, 7, and 8) that occur b' times; but all ten colors cannot occur this often, because $50 = 10 \times 5 = n \times b'$ exceeds $40 = 8 \times 5 = a' \times b'$, the area of the reduced rectangle.

Note that if *n* does divide *a* or *b*, it is trivial to cover the $a \times b$ rectangle with $1 \times n$ tiles, since all the tiles can be placed in the same orientation, with their sides of length *n* parallel to the side of the $a \times b$ rectangle which is divisible by *n*.

EXTENSION TO THE TORUS

If we have a rectangular board made of paper, we can make the left edge adjacent to the right edge by rolling it into a cylinder. If the original board were a sheet of rubber, we could not only turn it into a cylinder by bringing the left and right edges together, but we could then roll the cylinder into a torus (the mathematical name for a doughnut shape) by bringing the top and bottom edges together. We can think of a rectangular board as a torus without actually rolling it up at all if we decide that what disappears over the top reappears at the bottom, and that what goes over the right edge enters immediately at the left edge.

Early in 1986, I received a letter from Stan Wagon, telling me that he and Professor David Gale (University of California, Berkeley) were wondering whether de Bruijn's theorem would still be true if one regarded the $a \times b$ board as a *torus*. As a special case, they were considering the 10×10 "board," and 1×4 tiles. They thought the covering might be possible on the torus, and were considering writing a computer program to look for a solution. I quickly found the proof, embodied in figure 184, that in the special case of the 1×4 tile, de Bruijn's theorem is still true on the torus.

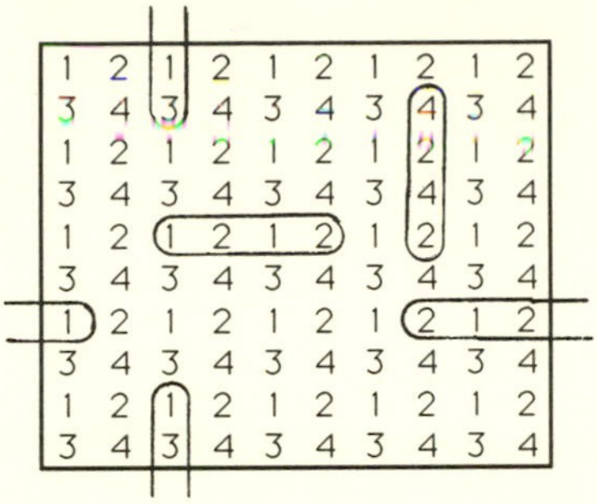

Figure 184. The 10×10 board, regarded as a torus, with 1×4 tiles.

We see that wherever the 1×4 tile is placed on the 10×10 toroidal board, it covers *an even number* of each of the four colors in figure 184. (Remember that *zero* is an even number!) Since the sum of even numbers is even, a *set* of 1×4 tiles must cover *even numbers* of each of the four colors. However, the 10×10 board has each color twenty-five times, and 25 is an *odd* number. Hence the entire board cannot be covered with 1×4 tiles, even when the board is regarded as a torus.

This proof immediately generalizes, for the 1×4 tile, to $a \times b$ toroidal boards, where a and b are both even but neither is a multiple of 4. The tiling is still impossible. More generally, if the tile is $1 \times k^2$, and the toroidal board is $a \times b$ where k divides both a and b, but k^2 divides neither a nor b, the tiling is impossible. (For this case, we use k^2 colors, arranged in $k \times k$ squares, following the same pattern as in figure 184, which illustrates the case $k = 2$.)

It is easy to verify that these impossibility proofs still work if we are allowed to place the $1 \times n$ tiles (where $n = k^2$) not only horizontally and vertically, but also "diagonally," on the $a \times b$ board! (When we place a $1 \times n$ tile "diagonally" at a 45-degree angle, we mean that it now covers n consecutive diagonal squares on the board. Thus it still has the same area, n, but no longer looks like a conventional rectangle.) For example, the 10×10 torus cannot be covered with 1×4 tiles, even if the tiles can be placed diagonally, as well as horizontally and vertically, on the torus. (This is still true if the diagonal has a slope different from 45 degrees, which has the effect of separating the squares of the tile in a uniform way.) "Uniformly spaced out" $1 \times n$ tiles can also be used horizontally or vertically without invalidating the impossibility proof. For example, using 1×4 tiles on the coloring in figure 184, all of these modified tiles still meet the essential requirement of covering an *even number* of each of the four colors.

I was then asked if I could prove that de Bruijn's theorem remains true on the torus with $1 \times n$ tiles on $a \times b$ "boards" for *all* values of n. I replied that I could not prove what was not true, and I produced the counterexample shown in figure 185. Specifically, the 10×15 torus *can* be tiled with 1×6 tiles, even though neither 10 nor 15 is a multiple of 6.

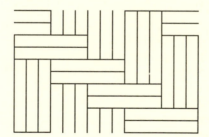

Figure 185. The 10 × 15 torus, covered with 1 × 6 tiles.

I also proved that this is the *smallest* example of an $a \times b$ torus that can be tiled with $1 \times n$ tiles, where n divides neither a nor b.

A tiling of the torus can also be regarded as a "doubly periodic tiling" of the infinite plane. For example, the pattern in figure 185 can be repeated periodically, to the right and left, and also above and below, to fill the entire plane with 1×6 tiles, but in such a way that the basic repeating unit is a 10×15 rectangle, even though the sides of any such rectangle will cut some of the 1×6 tiles. Such a floor-tiling pattern would be at least as interesting and esthetically satisfying as many that are currently in use.

Results similar to mine in figures 184 and 185 were obtained independently and at almost the same time by Professor Raphael M. Robinson (who was mentioned on page 8 in connection with the "triplication problem" for pentominoes).

ANOTHER FORMULATION OF DE BRUIJN'S THEOREM

The following is a slightly more general version of de Bruijn's theorem: "Suppose that an $X \times Y$ rectangle (where X and Y are real numbers) is exactly filled with rectangular tiles, where the tiles may have different sizes, but each tile has at least one side whose length is a whole number. Then either X or Y (or both) is a whole number." (We assume that the sides of the little rectangles are parallel to the sides of the big rectangle. Otherwise, they couldn't pack properly.)

Since this statement contemplates filling a large rectangle with a whole assortment of different-sized small rectangles, it obviously covers cases beyond the scope of de Bruijn's original statement. The first thing we will show is that de Bruijn's theorem really is included in this new statement.

Previously, we formulated de Bruijn's theorem in terms of an $a \times b$ rectangle to be filled with $1 \times n$ tiles, where a, b, and n are whole numbers. Let us adopt a new unit of measurement, n times larger than the old unit. (In the case of $n = 12$, this would correspond to measuring everything in *feet* rather than *inches*.) In terms of our new, larger units, the rectangle to be filled has dimensions $a/n \times b/n$, and the *tiles* are $1/n \times 1$. In particular, each tile has one side that is a whole number. The new form of de Bruijn's theorem then guarantees that at least one of the sides of the big rectangle (a/n or b/n) must be a whole number; that is, either a or b (or both) must be a multiple of n. That gives us our previous form of de Bruijn's theorem! (De Bruijn also treated $m \times n$ bricks in $a \times b$ rectangles.)

One of the first proofs given for this more general formulation of de Bruijn's theorem used a double integral of a complex exponential function: elegant, but certainly not elementary. In the spirit of this book, I will give a proof based on a checkerboard coloring. (The idea of this proof is due to Professors Richard H. Rochberg of Washington University and Sherman K. Stein of the University of California at Davis.)

Suppose we have the big $X \times Y$ rectangle completely filled with smaller rectangles, where each of the little rectangles has at least one side that is a *whole number* (that is, a multiple of *one unit*). We impose a checkerboard coloring, using black and white as the "colors" with the side of the basic checkerboard square equal to *one-half unit*, with the left and bottom sides of the big rectangle lined up with edge lines of the checkerboard pattern.

Since each of the little rectangles has at least one side that is a multiple of *one unit*, as we go along that direction we see that each little rectangle must cover an equal amount of white and black area! (Remember that the checkerboard pattern changes color every *half* of one unit.) Since this is true of each little rectangle, and the big rectangle is composed entirely of little rectangles, the big rectangle must also cover equal amounts of white and black area. It remains only to show that this implies that at least one side of the big rectangle must have a whole number as its length. We prove this by contradiction.

Suppose neither side of the big rectangle is a whole number. Let $X' = X - [X]$ and $Y' = Y - [Y]$, where $[X]$ denotes the largest whole number not exceeding X. (Thus X' and Y' are the *fractional parts* of X and Y, respectively.) As shown in figure 186, it is clear that except for the small corner rectangle of size $X' \times Y'$, the rest of the $X \times Y$ rectangle covers equal amounts of black and white area. We have assumed that $X' \neq 0$ and $Y' \neq 0$. We will show that in this case, the $X' \times Y'$ rectangle *cannot* cover equal amounts of black and white area.

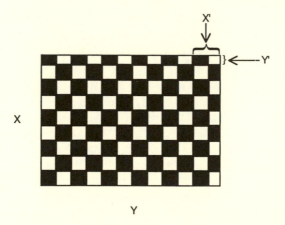

Figure 186. The half-unit checkerboard coloring applied to the $X \times Y$ rectangle.

We distinguish 4 cases, as illustrated in figure 187, depending on whether X' is less than or greater than $\frac{1}{2}$, and whether Y' is less than or greater than $\frac{1}{2}$. It is immediately obvious that the $X' \times Y'$ rectangle covers more black than white area, with the possible exception of the case $X' > \frac{1}{2}$, $Y' > \frac{1}{2}$. Even in

Figure 187. The four cases of the "residual rectangle."

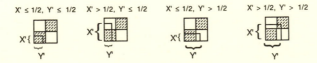

this case, there is more black than white, as we see by folding the figure first in one color midline and then in the other, and allowing opposite colors to cancel each other out. The net black area that remains in this case is exactly $(1 - X')(1 - Y')$, which is a positive area, since both factors are positive lengths. (This is illustrated in figure 188.) The proof by contradiction is now complete.

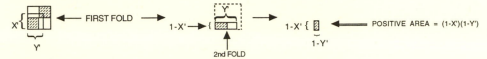

Figure 188. Folding the "residual rectangle" in case 4 to reveal a net excess of black over white area.

The similarity between this proof and the coloring proof given for the original form of de Bruijn's theorem (both proofs involve reduction to a "residual rectangle") is *not* coincidental.

RESULTS ABOUT POLYHEXES

The hexagonal animals (*polyhexes*, for short) are shown in figure 143 with up to five cells, and a 3-color periodic hexagonal tiling of the plane is shown in figure 142. The reader was invited to propose interesting configurations for the polyhexes. Here are several of my own.

The *hexagonal triangle of side 7*, shown in figure 189, consists of twenty-eight hexagons. Here are your tasks. (Solutions are given at the end of the chapter.)

Figure 189. The hexagonal triangle of side 7.

1. Tile the triangle in figure 189 with seven copies of a single tetrahex (4-celled hexagonal animal) shape. (Rotating the shape and turning it over are permitted.)
2. Can this triangle in figure 189 be tiled with any combination of the shapes

    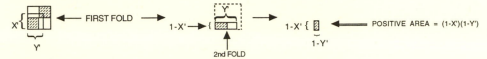 , , and ?

3. Can the triangle be tiled with any combination of the shapes

    , , and ?

4. Is it possible to tile the triangle using exactly *one of each* of the seven tetrahex shapes?

Chapter 9

5. Using exactly one of each of the 7 tetrahex shapes, is it possible to tile either (or both) of the 4 × 7 hexagonal "rectangles" shown in figure 190?

Figure 190.
The two 4 × 7
hexagonal
"rectangles."

 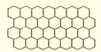

6. Figure 191 shows the hexagonal triangles of sides 2 through 6. (The triangle of side 7 was shown in figure 189.) Is it possible to tile any larger hexagonal triangle entirely with copies of the triangle of side 2?

Figure 191.
The hexagonal
triangles of sides
2 through 6.

7. The hexagonal triangle of side 8 consists of thirty-six hexagons. Can it be tiled with twelve trihexes, if exactly one of them is ⬡⬡ , and the other eleven are any combination of ⬡⬡⬡ and ?

8. The hexagonal triangle of side 11 consists of sixty-six hexagons. Can it be tiled with twenty-two copies of the "straight trihex" ⬡⬡⬡ ?

SOLUTIONS TO POLYHEX PROBLEMS

1. Only one of the seven tetrahex shapes can be used to tile the hexagonal triangle of side 7. The solution is shown in figure 192.

2. There are two solutions, as shown in figure 193. One has central symmetry; the other does not.

Figure 192. The
unique tetrahex
shape that tiles the
hexagonal triangle
of side 7.

Figure 193. The two solutions to problem 2.

3. No such tiling is possible. To prove this, color the triangle in four colors, as illustrated in figure 194. Each

of the three tetrahex shapes in this problem will cover one cell of each color, no matter where it is placed on the triangle. (Convince yourself of this!) Hence, a tiling with these shapes must cover *equally many cells* of each of the four colors. However, by actual count, the triangle contains *ten* "a" cells, but only 6 cells of each of the other three colors.

Figure 194. *A periodic coloring of the hexagonal tiling in four colors.*

4. No such tiling exists. One way to show this is to start with the three inequivalent locations of the "propeller," shown in figure 195. Each of these starts can be searched exhaustively (tediously by hand, or more quickly by computer) to show that no successful completion of the tiling, using one of each of the seven tetrahexes, is possible. (Did anyone find a simpler proof of impossibility?)

Figure 195. *The three inequivalent locations for the "propeller" on the hexagonal triangle of side 7.*

5. Solutions to both "rectangles" are shown in figure 196.

Figure 196. *Tiling each of the 4 × 7 "rectangles" with a set of seven tetrahexes.*

6. Yes. The first two examples are the hexagonal triangles of side 9 and side 11, as shown in figure 197.

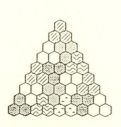 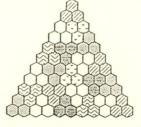

Figure 197. *Successful tilings of large hexagonal triangles with copies of the hexagonal triangle of side 2.*

Chapter 9

7. No such tiling is possible. To see this, color the side-8 triangle in three colors, as shown in figure 198. In this coloring, each of the three colors occurs twelve times.

Figure 198.
A 3-coloring for
hexagonal triangles.

Wherever it is placed, the "straight trihex" ⬡⬡⬡ will cover one cell of each color. The triangle of side 2,

⬡⬡⬡ , will also cover one cell of each color, no matter where it is placed. However, the "chevron" ⬡⬡⬡

will cover two cells of one color, one cell of a second color, and no cells of the third color. Thus, a single "chevron" will destroy the balance in the occurrence of the three colors.

8. This is another impossible tiling. Our proof uses the numbering pattern shown in figure 199. A "straight trihex" placed anywhere covers three cells whose numbers sum to either 0 or 3 or 6. Thus, any set of straight trihexes covers a set of cells whose numbers sum to a multiple of 3. However, the total over all sixty-six cells of the triangle is 62, which is *not* a multiple of 3, so the tiling is impossible. (*Note:* This proof idea can be used to show the impossibility of tiling a hexagonal triangle of side n with straight trihexes except when n is a multiple of 9, or one less than a multiple of 9. An impossibility proof for *all* values of n was given in 1990 by J. H. Conway and J. C. Lagarias, using principles from finite group theory.)

Figure 199.
Another 3-coloring
for hexagonal
triangles.

Answers to Exercises
in Chapter 5

1. $8^2 \times 10^5 = 6{,}400{,}000$.

2. $12 \times 15 \times 9 \times 13 \times 11 \times 12 = 2{,}779{,}920$.

3. $26 \cdot 25 \cdot 24 - 1 = 15{,}599$.

4. $52 \cdot 51 \cdot 50 \cdot 49 \cdot 48 \cdot 47 = 14{,}658{,}134{,}400$.

5. $\binom{64}{4} = \dfrac{64 \cdot 63 \cdot 62 \cdot 61}{4 \cdot 3 \cdot 2 \cdot 1} = 635{,}376$.

6. $\binom{64}{8} = \dfrac{64 \cdot 63 \cdot 62 \cdot 61 \cdot 60 \cdot 59 \cdot 58 \cdot 57}{8 \cdot 7 \cdot 6 \cdot 5 \cdot 4 \cdot 3 \cdot 2 \cdot 1} = 4{,}426{,}165{,}368$.

7. $\binom{12}{6} = \dfrac{12 \cdot 11 \cdot 10 \cdot 9 \cdot 8 \cdot 7}{6 \cdot 5 \cdot 4 \cdot 3 \cdot 2 \cdot 1} = 924$.

8. $\binom{7}{3} = \dfrac{7 \cdot 6 \cdot 5}{3 \cdot 2 \cdot 1} = 35$.

9. $\binom{4 + 5 - 1}{5 - 1} = \binom{8}{4} = 70$.

1111	1112	4445	2235	1355
2222	1113	1555	2245	1455
3333	1114	2555	1233	2355
4444	1115	3555	1334	2455
5555	1222	4555	1335	3455
1122	2223	1123	2334	1234
1133	2224	1124	2335	1235
1144	2225	1125	3345	1245
1155	1333	1134	1244	1345
2233	2333	1135	1344	2345
2244	3334	1145	2344	
2255	3335	1223	1445	
3344	1444	1224	2445	
3355	2444	1225	3445	
4455	3444	2234	1255	

10. $\binom{3 + 12 - 1}{3 - 1} = \binom{14}{2} = 91$.

11. If there were extra marks on the symbols to make them all distinguishable, there would be $k!$ permutations. However, the k_1 symbols of the first type can be permuted among themselves in $k_1!$ ways, and the k_2 symbols of the second type can be independently permuted among themselves in $k_2!$ ways, and so on, leaving only $\dfrac{k!}{k_1! \, k_2! \, \ldots \, k_r!}$ distinguishable permutations.

12. For PEPPER, $\dfrac{6!}{3! \, 2! \, 1!} = 60$ permutations. For MISSISSIPPI, $\dfrac{11!}{4! \, 4! \, 2! \, 1!} = 34{,}650$ permutations.

13.

n	No Factors in Common with n	$\phi(n)$
1	1	1
2	1	1
3	1, 2	2
4	1, 3	2
5	1, 2, 3, 4	4
6	1, 5	2
7	1, 2, 3, 4, 5, 6	6
8	1, 3, 5, 7	4
9	1, 2, 4, 5, 7, 8	6
10	1, 3, 7, 9	4
11	1, 2, 3, 4, 5, 6, 7, 8, 9, 10	10
12	1, 5, 7, 11	4
13	1, 2, 3, 4, 5, 6, 7, 8, 9, 10, 11, 12	12
14	1, 3, 5, 9, 11, 13	6
15	1, 2, 4, 7, 8, 11, 13, 14	8

14. The derivation of the formula is indicated in the "Hints." The rest is left to the reader.

15. The primes to 100 are: 2, 3, 5, 7, 11, 13, 17, 19, 23, 29, 31, 37, 41, 43, 47, 53, 59, 61, 67, 71, 73, 79, 83, 89, 97. By the formula,

$$\pi(100) = 4 - 1 + 100 - \left(\left[\frac{100}{2} \right] + \left[\frac{100}{3} \right] + \left[\frac{100}{5} \right] \right.$$

$$\left. + \left[\frac{100}{7} \right] \right) + \left(\left[\frac{100}{6} \right] + \left[\frac{100}{10} \right] + \left[\frac{100}{14} \right] \right.$$

$$\left. + \left[\frac{100}{15} \right] + \left[\frac{100}{21} \right] + \left[\frac{100}{35} \right] \right) - \left(\left[\frac{100}{30} \right] \right.$$

$$+ \left[\frac{100}{42}\right] + \left[\frac{100}{70}\right] + \left[\frac{100}{105}\right]\right) + \left[\frac{100}{210}\right]$$

$$= 103 - (50 + 33 + 20 + 14) + (16 + 10 + 7$$

$$+ 6 + 4 + 2) - (3 + 2 + 1 + 0) + 0 = 103$$

$$- 117 + 45 - 6 = 25.$$

16. Eighteen were healthy.

17. $\frac{1}{2}(T + C) = \frac{1}{2}(1000 + 100) = 550.$

18. $\frac{1}{2}(T + C) = \frac{1}{2}(26^4 + 26^2) = 228{,}826.$

19. When k is an even number, $N = \frac{1}{2}(n^k + n^{k/2}).$
    When k is an odd number, $N = \frac{1}{2}(n^k + n^{(k+1)/2}).$

20. Thirty-two.

21. Twenty-eight double-6 and fifty-five double-9 dominoes.

22. The twenty-seven drawings will not be presented here.

23. $N = \frac{1}{4}[n^4 + 3n^2].$

24. $N = \frac{1}{4}(T + C_a + C_b + C_c) = \frac{1}{4}(5^6 + 5^3 + 5^3 + 5^4) = 4125.$

25. H, I, O, X.

26. With a and b both odd: $N = \frac{1}{4}(T + C_a + C_b + C_c) = \frac{1}{4}(2^{ab} + 2^{a(b+1)/2} + 2^{((a+1)/2)b} + 2^{(ab+1)/2})$; a and b both even: $N = \frac{1}{4}(T + C_a + C_b + C_c) = \frac{1}{4}(2^{ab} + 2^{ab/2} + 2^{ab/2} + 2^{ab/2}) = \frac{1}{4}(2^{ab} + 3 \cdot 2^{ab/2})$; a odd, b even: $N = \frac{1}{4}(T + C_a + C_b + C_c) = \frac{1}{4}(2^{ab} + 2^{ab/2} + 2^{((a+1)/2)b} + 2^{ab/2}).$

27. $N = \frac{1}{4}(3^4 + 2 \cdot 3 + 3^2) = 24.$

28. Twenty-four, as before. One rectangular pattern is

29. $N_4 = \frac{1}{4}(4^4 + 2 \cdot 4 + 4^2) = 70.$ $N_n = \frac{1}{4}(n^4 + 2n + n^2) = \frac{1}{4}n(n + 1)(n^2 - n + 2).$ If n or $n + 1$ is a multiple of 4, this is clearly an integer. However, in any case, one of n and $n + 1$ is even, and $n^2 - n + 2$ is always even, so that the product $n(n + 1)(n^2 - n + 2)$ is always a multiple of 4.

30. $N_2 = \frac{1}{4}(2^9 + 2 \cdot 2^3 + 2^5) = 140.$

31. $N_n = \frac{1}{4}(n^9 + 2 \cdot n^3 + n^5)$.

32. For even k, $N_n = \frac{1}{4}(n^{k^2} + 2 \cdot n^{k^2/4} + n^{k^2/2})$. For odd k, $N_n = \frac{1}{4}(n^{k^2} + 2 \cdot n^{(k^2+3)/4} + n^{(k^2+1)/2})$.

33. An octomino symmetric under 90-degree rotation:

$$N_2 = \frac{1}{4}(2^8 + 2 \cdot 2^2 + 2^4) = 70.$$

34. $N = \frac{1}{4}(n^4 + 2 \cdot n + n^2)$.

35. $N = \frac{1}{8}(T + C_{180} + 2C_H + 2C_D) = \frac{1}{8}\left[\binom{9}{3} + 4 + 2 \cdot 10 + 2 \cdot 10\right] = 16.$

Three monominoes can be placed on a 3×3 board in the sixteen following ways (each monomino is indicated by a dot):

36. For the 4×4 board:

$$N_4 = \frac{1}{8}(T + C_{180} + 2C_{90} + 2C_H + 2C_D) = \frac{1}{8}\left\{\binom{16}{4} + \binom{8}{2} + 2 \cdot 4 + 2 \cdot \binom{8}{2} + 2 \cdot (1 + 6 \cdot 6 + 15)\right\} = 252.$$

For the 6×6 board:

$$N_6 = \frac{1}{8}\left\{\binom{36}{4} + \binom{18}{2} + 2 \cdot 9 + 2 \cdot \binom{18}{2} + 2\left[\binom{6}{4} + \binom{6}{2}\binom{15}{1} + \binom{15}{2}\right]\right\} = 7,509.$$

For the 8×8 board:

$$N_8 = \frac{1}{8}\left\{\binom{64}{4} + \binom{32}{2} + 2 \cdot 16 + 2 \cdot \binom{32}{2} + 2\left[\binom{8}{4} + \binom{8}{2}\binom{28}{1} + \binom{28}{2}\right]\right\} = 79,920.$$

37. $N = \frac{1}{8}(3^{16} + 3^8 + 2 \cdot 3^4 + 2 \cdot 3^8 + 2 \cdot 3^{10}) = 5,398,083.$

38. $N = \frac{1}{8}\left[\binom{9}{5} + \binom{4}{2} + 2\cdot 2 + 2\cdot 12 + 2\cdot 12\right] = 23.$

39. $T = 3! \binom{8}{3}^2 = 18{,}816.$

$N = \frac{1}{8}\left[18{,}816 + 0 + 2\cdot 0 + 2\cdot 0 + 2\cdot 4\cdot\binom{8}{3}\right] = 2{,}408.$

40. For one monomino: $N_1 = \dfrac{n^2 + 2n}{8}$ for even n;

$N_1 = \dfrac{n^2 + 4n + 3}{8}$ for odd n.

For two monominoes: $N_2 = \frac{1}{16}(n^4 + 6n^2 - 4n)$ for even n; $N_2 = \frac{1}{16}(n^4 + 8n^2 - 8n - 1)$ for odd n.

41. The X pentomino has the symmetry group of the square. The I pentomino has that of the rectangle. The T, U, V, W, and Z pentominoes have involutional symmetry. The F, L, N, P, and Y pentominoes have only identity symmetries.

42. $N_3 = \frac{1}{8}(3^5 + 3^3 + 2\cdot 3^2 + 2\cdot 3^4 + 2\cdot 3^3) = 63$ altogether.

$N_{abc} = N_3 - 3N_2 + 3N_1 = 63 - 36 + 9 = 36$ ways that actually use all three colors.

43. Without regard for symmetries: $T_4 = 4\cdot 3\cdot 2\cdot 1 = 24$. For distinguishable cases under rotation and reflection:

$N_4 = \frac{1}{8}(24 + 8 + 2\cdot 2 + 0 + 2\cdot 10) = 7.$

44. For six rooks: $T_6 = 6! = 720.$

$N_6 = \frac{1}{8}\left\{720 + 6\cdot 4\cdot 2 + 0 + 0 + 2\left[1 + \binom{6}{4} + \binom{6}{2}\cdot 3 + 15\right]\right\} = 115.$

For eight rooks: $T_8 = 8! = 40{,}320.$

$N_8 = \frac{1}{8}\left\{8! + (8\cdot 6\cdot 4\cdot 2) + 2\cdot 12 + 0 + 2\left[1 + \binom{8}{6}\cdot 1 + \binom{8}{4}\cdot 3\cdot 1 + \binom{8}{2}\cdot 5\cdot 3\cdot 1 + \binom{8}{0}\cdot 7\cdot 5\cdot 3\cdot 1\right]\right\} = 5{,}282.$

45. $N = \frac{1}{6}(2^6 + 2 + 2^2 + 2^3 + 2^2 + 2) = 14.$

46. $N = \frac{1}{12}(2^6 + 2 + 2^2 + 2^3 + 2^2 + 2 + 3\cdot 2^4 + 3\cdot 2^3) = 13.$

47. With rotations only, and k colors, $N = \frac{1}{6}(k^6 + k^3 + 2k^2 + 2k).$

With rotations and reflections, and k colors, $N = \frac{1}{12}(k^6 + k^3 + 2k^2 + 2k + 3k^4 + 3k^3) = \frac{1}{12}(k^6 + 3k^4 + 4k^3 + 2k^2 + 2k)$. In particular, with rotations only, $N_3 = 130$ and $N_4 = 700$, while with rotations and reflections, $N_3 = 92$ and $N_4 = 430.$

48. $N = \frac{1}{6}(5^3 + 2\cdot 5 + 3\cdot 5^2) = 35.$

49. The number of distinct strings is b^p. For the necklaces with the cyclic group of symmetries, $N = \frac{1}{p}[b^p + (p - 1)b]$; with the dihedral group of symmetries, $N = \frac{1}{2p}[b^p + (p - 1)b + pb^{(p+1)/2}]$.

For $b = 2$ and $p = 5$, both enumerations yield an answer of 8. For $b = 4$ and $p = 3$, these numbers are 24 and 20, respectively.

50. The twenty-four symmetries are: one identity; six rotations by ± 90 degrees about a face-to-face diagonal; three rotations by 180 degrees about a face-to-face diagonal; six rotations by 180 degrees about an edge-to-edge diagonal; eight rotations by ± 120 degrees about a vertex-to-vertex diagonal.

51. $N = \frac{1}{24}(6! + 0 + 0 + 0 + 0) = 30$.

52. $N = 2$.

53. $N = 6$.

54. When the number of labeled vertices is unrestricted.

$N = \frac{1}{24}(2^8 + 6 \cdot 2^2 + 3 \cdot 2^4 + 6 \cdot 2^4 + 8 \cdot 2^4) = 23$.

When there are four labeled and four unlabeled vertices.

$N = \frac{1}{24}\left[\binom{8}{4} + 6 \cdot 2 + 3 \cdot 6 + 6 \cdot 6 + 8 \cdot 2^2\right] = 7$.

Problem Compendium

THIS compendium contains all the problems dealing with the fitting together of pentominoes and related polyominoes that are included in the book as well as a number of additional constructions.

PENTOMINO PROBLEMS

The patterns in problems 1 through 32 can be covered exactly, using the twelve pentominoes. When an interesting special case is known (for example, construction from two identical pieces), it is indicated by a heavy line dividing the pattern.

1. Fit the twelve pentominoes into one 3×20 rectangle.
2. Again, using the twelve pentominoes, fit them into one 4×15 rectangle.
3. All twelve pentominoes can be fitted into one 5×12 rectangle.
4. Fit the pentominoes into one 6×10 rectangle.

5.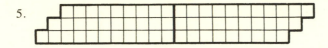

6.

7.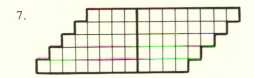

8.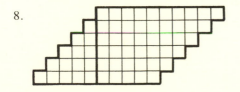

Appendix B

9.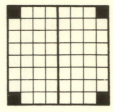

10.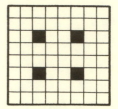

11.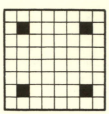

12.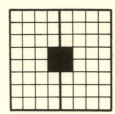

13.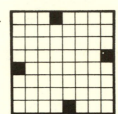

14.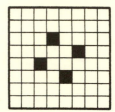

15.

16.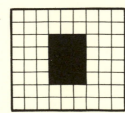

17.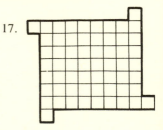

18.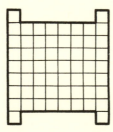

19.

20. Each of the three rectangles has one hole.

 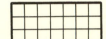 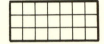

21. This 61-square pattern has a hole that, as was shown in chapter 4, cannot be located in the center.

22. If the 8 × 8 square is constructed from the two congruent pieces as shown in the diagram on the left, the unshaded half may be shifted bodily to form the 9 × 7 rectangle shown on the right.

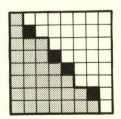 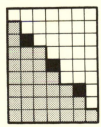

23. An unidentified reader of the *Fairy Chess Review* designed and solved this unusual problem. The twelve pentominoes will cover the irregular shape shown below, which can then be folded on the dashed lines to cover the surface of a cube as indicated.

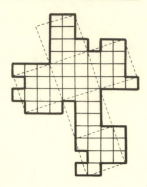

24.

With 2,339 solutions to the basic 6 × 10 rectangular pattern, additional constraints may be imposed to make the problem more challenging. Some of the more interesting special cases of this pattern are given in problems 25 through 32.

25. Construct two 6 × 5 rectangles.

26. The 6 × 10 rectangle contains a 3 × 5 subrectangle within it.

27. The 6 × 10 rectangle is constructed with a 4 × 5 subrectangle within it.

28. The 6 × 10 rectangle is constructed from the two pieces shown at the left. The unshaded half may be shifted as a unit to form the 9 × 7 rectangle shown at the right.

 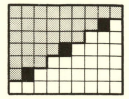

29. In a way similar to Problem 28, the unshaded portion of the 6 × 10 rectangle is shifted to form the 9 × 7 rectangle.

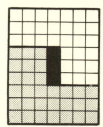

30. Build the 6 × 10 rectangle so that each of the twelve pentominoes touches an edge.

31. Build the 6 × 10 rectangle so that the I pentomino does not touch an edge.

32. Throw the twelve pentominoes randomly on the table. Now construct the 6 × 10 rectangle without turning over any of the pieces. (All thirty-two cases are possible.)

33. *The 10 Problem.* (*a*) Build a 10-square pattern with two of the pentominoes. Find two other pentominoes that will fill the same shape. (*b*) Use four of the remaining pieces to repeat the process. (*c*) Finally, use the last four pentominoes to repeat the process a third time. (The 10-square pattern of *b* need not be the same as the first, and the 10-square pattern of *c* may be different from both the first and second.)

34. *The 20 Problem.* Use four of the pentominoes to build a 20-square pattern. Use four others to build the *same* pattern again. The same pattern must now be constructed a third time with the last four pieces.

35. *The Double-Duplication Problem.* Use two pentominoes to build a 10-square pattern. Use two more pieces to repeat the shape. Use the remaining eight pieces to build a model of the shape, but with each of the linear dimensions multiplied by 2. The figure below gives an example of one of the several shapes that can be constructed in this manner.

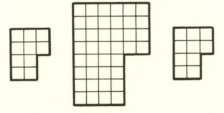

36. Figure 19 shows how the twelve pentominoes may be cut from a 6 × 13 rectangular piece of wood with a saw that does not cut around corners. Alison Doig, of London, England, has shown how this problem may be solved using either a 4 × 19 or a 5 × 15 piece of wood. The reader is invited to match Doig's effort and to dissect the 5 × 15 rectangle below. (The U pentomino again requires special attention. Assume that it must be cut as a 2 × 3 rectangle and finished subsequently, by other means.)

37. *The Triplication Problem.* Given a pentomino, use nine of the other pentominoes to construct a scale model three times as wide and three times as high as the given piece. Construct all twelve pentomino triplications.

38. An even number of P pentominoes can easily cover a rectangle with an even area. Find the smallest rectangle that can be covered with an *odd* number of P pentominoes.

39. Make a 9 × 10 rectangle with the eighteen one-sided pentominoes shown below. These pieces may not be turned over.

SOLID PENTOMINO PROBLEMS

Problems 40 through 46 can be constructed using a set of solid pentominoes, that is, the twelve pentominoes, each constructed from five "unit cubes."

40. Construct a 3 × 4 × 5 solid.

41. Build a 2 × 5 × 6 solid. (This can be constructed from two 1 × 5 × 6 planar subunits.)

42. Build a 2 × 3 × 10 solid. (This cannot be constructed from two 1 × 3 × 10 subunits. However, a "minimum" solution, with only the L and Y pieces not wholly contained in one of the 1 × 3 × 10 planes, does exist.)

43. The Cinder Block (the shaded area is empty).

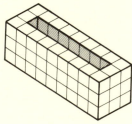

44. The Steps.

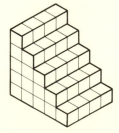

45. The Pyramid. Only eleven of the pieces are used for this shape, but it still ranks with the more difficult problems.

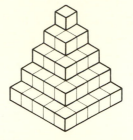

46. *Model Problems.* A solid model of a solid pentomino can be made with a volume of sixty cubes. The scale model is twice the length and twice the width of the solid pentomino, and is built three units deep. The figures below show the models of the T and U pentominoes constructed in such a fashion. The model of the U is particularly interesting since it can be inverted, as shown, and called a tunnel or arch. Of the twelve pentominoes, the I, L, P, N, T, U, V, Y, and Z are known to have solutions, the W and X are impossible, and whether or not the F has a solution is still undecided.

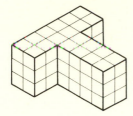 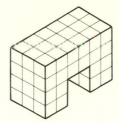

PENTACUBE PROBLEMS

Pentacubes, as noted earlier, use all solid pentominoes that are distinct under 3-dimensional rotation. A complete set consists of all possible combinations of five cubes connected on their faces. Problems 47 through 57 were all designed and solved by David A. Klarner. All the problems using twenty-eight pieces can be solved with the "straight" pentacube omitted.

47. Construct a set of the twenty-nine pentacubes and identify the six mirror-image pairs.

48. Use twenty-eight of the pieces to construct a 2 × 5 × 14 solid.

49. Use twenty-eight of the pieces to construct a 2 × 7 × 10 solid.

50. Use twenty-eight of the pieces to construct a 4 × 5 × 7 solid.

51. Find a simultaneous solution to the last three problems by constructing two solids, each 2 × 5 × 7.

52. Construct the $2 \times 5 \times 14$ solid from five smaller rectangular solids.

53. Use twenty-five pieces to build a 5-unit cube. The four pieces omitted may be the four that have a linear dimension of four units or more.

54. Build a 7-layer pyramid. It is constructed like problem 45 but with two more layers added to the bottom (a 6×6 and a 7×7 layer).

55. Use twenty-seven of the pentacubes to build a model of a given pentacube. Build such models of all twenty-nine pentacubes.

56. Use twenty-eight of the pentacubes to build a cylinder seven units high with its cross section a model of one of the solid pentominoes. All twelve problems are possible.

57. *The Checkerboard*. Use eighteen of the pentacubes to make a $3 \times 6 \times 6$ solid with alternate cubes missing from one of the 6×6 faces.

OTHER POLYOMINO PROBLEMS

Problems 58 through 61 use a combined set of the twelve pentominoes and the 5 tetrominoes.

58. Construct an 8×10 rectangle.

59. Using all the above pieces, build a 4×20 rectangle.

60. Obtain a simultaneous solution to the previous two problems by constructing two 4×10 rectangles.

61. Construct a 5×16 rectangle.

Problems 62–67. Use a set of the thirty-five hexominoes to build the patterns that follow. (These constructions first appeared in *Fairy Chess Review*.)

62.

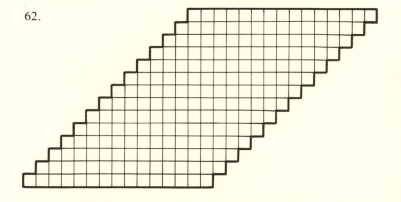

63.

64.

65.

66.

67.

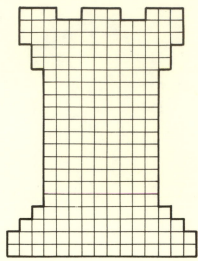

68. Use the thirty-five hexominoes and the twelve pentominoes to build an 18 × 15 rectangle. The special case indicated below has the pentominoes forming a "rook" in the center of the rectangle. (This problem also appeared in *Fairy Chess Review*.)

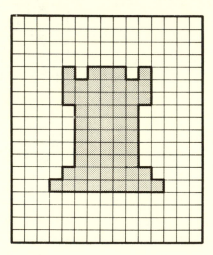

69. Use the 108 heptominoes to build three 11 × 23 rectangles each with a hole in the center (see fig. 103).

70. Discard the heptomino with the hole, and build a 107 × 7 rectangle with the remaining 107 heptominoes. This may be done by building four 7 × 25 rectangles and one 7 × 7 square (see fig. 104).

IMPOSSIBLE CONFIGURATIONS

The following problems all involve regions that the twelve pentominoes will not cover. For the asterisked problems, no simple impossibility proof has yet been discovered.

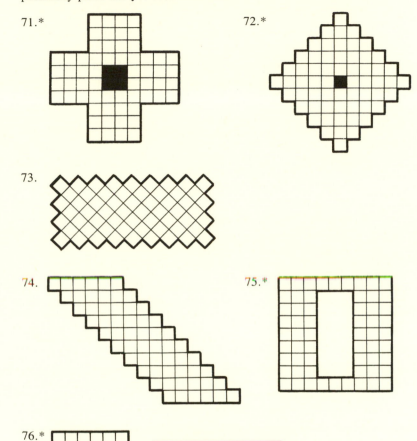

71.* 72.*

73.

74. 75.*

76.*

77.*Impossible with the twelve solid pentominoes.

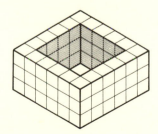

78. Impossible with the twelve solid pentominoes.

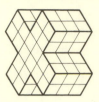

79.*The 3 × 20 rectangle constructed with the twelve pentominoes cannot be made from two smaller rectangles.

80. Place four monominoes on the 8 × 8 checkerboard without completely isolating any region of only a few squares, but in such a way that the twelve pentominoes will not cover the remainder of the board.

READERS' RESEARCH

The following pentomino problems were not known to have solutions, nor had they been proved impossible, in 1965. (For their current status, see Appendix C.)

81.   82.

83.

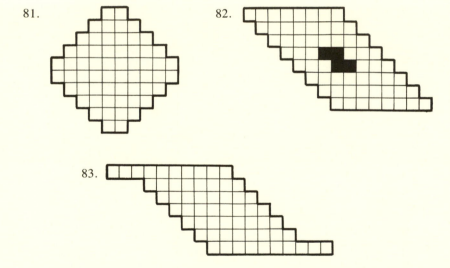

Problems 24 and 25 are the only known solutions involving simultaneous rectangles (that is, two rectangles made out of the twelve pentominoes with no pentomino used more than once) and pentominoes. The only other results known on this topic are that simultaneous 3 × 10 and 6 × 5 rectangles cannot be constructed (problem 76); the 3 × 20 rectangle cannot be constructed from two smaller rectangles (problem 79); and a 2 × 10 and 4 × 10 rectangle cannot be constructed simultaneously. Problems 84 through 87 give other rectangular factorizations that might be possible.

84. Construct 4 × 5 and 5 × 8 simultaneous rectangles.

85. Build simultaneous 3 × 5 and 5 × 9 rectangles.

86. Simultaneous 4 × 5 and 4 × 10 rectangles are to be constructed.

87. Construct 2 × 10 and 5 × 8 simultaneous rectangles.

88. *The 15 Problem.* Can a pattern of fifteen squares be found that can be constructed four times simultaneously from a set of the pentominoes? (*Cf.* problems 33 and 34.)

89. Many variations on the 3 × 20 rectangle involve configurations with a 3-unit width. The cross is typical of these possibilities.

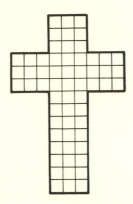

90. Another variation of a 3 × 20 rectangle, with a hole in the center:

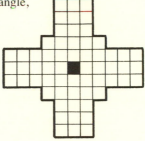

91. It is known that nine of the pentominoes can be used to triplicate any given pentomino (see problem 37). Can nine sets be used to triplicate the complete set of twelve pentominoes?

92. Use the solid pentominoes to build the model of the F.

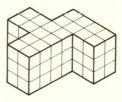

Undoubtedly readers can invent many new polyomino and polycube problems for themselves. There is also a great deal of unexplored problem territory based on the material in chapter 7.

C Golomb's Twelve Pentomino Problems

by Andy Liu

1. A BRIEF SURVEY

The polyominoes are geometric shapes formed of unit squares joined edge-to-edge. There are one monomino, one domino (from which the "family name" is derived), two trominoes, and five tetrominoes. The most attractive are the 12 pentominoes, shown in figure C.1 with their "letter names."

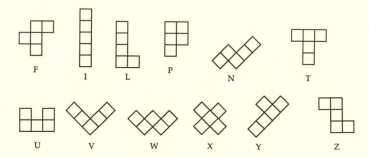

*Figure 1.
Golomb's names
for the twelve
pentominoes.*

Their number is small enough to be manageable and large enough to offer diversity. They were first featured in a classic problem of Dudeney [3], then discussed by Dawson and Lester [2], and popularized by Golomb [10] and especially Martin Gardner [7, 8, 9].

In his masterpiece, *Polyominoes,* Golomb [11] discussed many fascinating aspects of these combinatorial pieces. To whet the appetite of enthusiasts, he posed twelve "readers' research problems." More than twenty years after the publication of this classic, one of these problems is still unsolved.

In each of problems 84 to 87, the readers are asked to construct a pair of rectangles using a complete set of pentominoes, or prove that no such constructions are possible. The rectangles are 4 × 5 and 5 × 8 in problem 84, 3 × 5 and 5 × 9 in problem 85, 4 × 5 and 4 × 10 in problem 86, and 2 × 10 and 5 × 8 in problem 87. In each of problems 81, 82, 83, 89, and 90, a specific shape is to be constructed. These are shown in figure C.2.

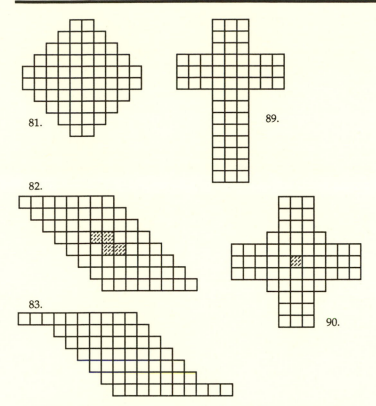

Figure 2.
Golomb's problems
81, 82, 83, 89,
and 90.

Problems 82 and 83 were solved by Walsh [17], problems 89 and 90 by Garcia [5], and problem 85 by Fairbairn [4]. These solutions are all in the affirmative, and unpublished [6].

Fairbairn's solution shows up in an exhaustive list he had compiled of 1010 ways of constructing a 5 × 12 rectangle. Haselgrove and Haselgrove [12] had an earlier computer program that generated an exhaustive list of 2339 ways of constructing a 6 × 10 rectangle. Miller [15] pointed out that it could have been run for a 5 × 12 rectangle as well, but this was not done.

Fairbairn's list contains no construction that would solve problem 84. It may be desirable to obtain a proof of impossibility without reference to such lists.

Solutions to problems 82, 83, 85, 89 and 90, found independently, were published by Liu [14], whose article also contained a proof that the simultaneous rectangles in problem 87 cannot be constructed. An affirmative solution to problem 86 was found by Bouwkamp [1]. Here I shall prove that the construction required in problem 81 is not possible.

These nine "standard" pentomino problems share the following characteristics:

1. The construction of a specific shape is sought.
2. Only one complete set of pentominoes is used.
3. The pentominoes are considered planar figures.

Appendix C

In each of the remaining three problems exactly one of the above is suppressed.

A problem in which (1) is suppressed simply asks the readers to use a complete planar set of pentominoes to construct an unspecified shape. Clearly a modification is in order. In problem 88, the only one of the twelve unsolved, readers are asked to choose a shape consisting of fifteen unit squares and construct four copies of it.

In problem 91, the readers are asked to use nine complete sets of pentominoes to construct a set of models of the pentominoes three times as long and as wide. Implicit are the requirements that the nine pieces used to construct each model must not contain duplicates, and must not contain the piece corresponding to the model.

Figure C.3 shows one such construction, independent of prior unpublished solutions obtained by Wade Philpott [13] and by several people in Japan [18], some by hand and some with the assistance of computers. Figure C.3 was obtained by hand.

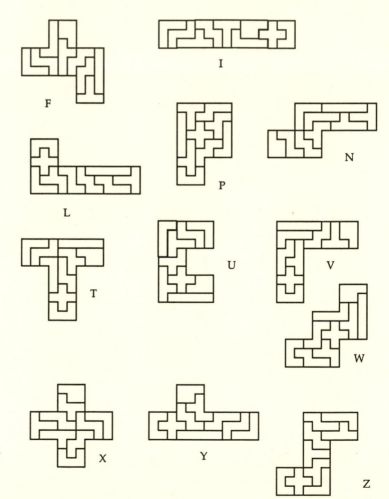

Figure 3. The pentominoes formed with nine sets of pentominoes.

In problem 92, the pentominoes are considered solid figures with unit thickness, and the construction of a model of the F-pentomino is sought, where the model is twice as long and as wide but three times as thick. A solution was published by Verbakel [16].

2. SOLUTION TO PROBLEM 81

We shall prove that the desired shape, featured in figures C.4 and C.5, cannot be constructed with a complete set of pentominoes.

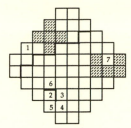

Figure 4.

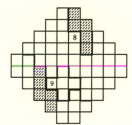

Figure 5.

There are twenty cells on the boundary of the shape. The maximum number of these cells that can be covered by a particular pentomino is as follows:

F	I	L	N	P	T	U	V	W	X	Y	Z
3	1	2	2	3	1	2	1	3	2	2	1

The total is 23, so there is no immediate contradiction. However, these boundary cells will play an important role in our argument. If the X pentomino is to occupy its maximum quota of boundary cells, or in fact any boundary cells, it must be placed, up to rotation and reflexion, as shown in the upper part of figure C.4. Note that the top part of the shape must now be occupied by the P pentomino.

Consider cell number 1. It cannot be occupied by the I pentomino. If it is occupied by any of the T, U, V, or Y pentominoes, regions in which no pentomino can fit will be isolated. If it is occupied by any of the F, L, N, or Z pentominoes without isolating impossible regions, then the left-most part of the shape must be occupied by the P pentomino, which, however, had been committed.

It now follows that cell number 1 must be occupied by the W pentomino. The orientation shown is the only possible one since al-

Appendix C

ternative placements of the W pentomino either isolate an impossible region or necessitate the use of another copy of the P pentomino.

Consider the cells numbered 2 and 3. We claim that they cannot be occupied by the same pentomino. Suppose the contrary is true. Then such a pentomino cannot also occupy cell number 4. Otherwise, it must occupy cell number 5 as well, but the P pentomino is the only one that can cover a 2×2 square. Now the bottom part of the shape must be occupied by the P pentomino, which is no longer available.

It follows that the pentomino occupying cell number 2 cannot occupy cell number 3, nor can it, by symmetry, occupy cell number 6. Hence it must be one of the L, N, U, W, and Z pentominoes. However, all except the W pentomino will isolate impossible regions, and the W pentomino has been committed. Hence the X pentomino cannot occupy any boundary cell.

This means that exactly one other pentomino may occupy one less than its maximum quota of boundary cells while the others must occupy their maximum quota. A pentomino other than the X pentomino is said to be *deficient* if it occupies less than its maximum quota.

Suppose the U pentomino is not deficient. Then it must be placed, again up to rotation and reflexion, as shown in the right-most part of figure C.4. Now cell number 7 can be occupied only by one of the F, L, and W pentominoes, which will then be deficient. It follows that either the U pentomino is deficient or one of the F, L, and W pentominoes is deficient, and in the manner described.

In any case, the N pentomino is not deficient. It has two nonequivalent nondeficient placements, as shown in the upper and the lower parts of figure C.5. In the former case, cell number 8 cannot be occupied without creating multiple deficiency when the U pentomino is taken into consideration.

Finally, in the latter case, cell number 9 can be occupied nondeficiently only by the X pentomino as shown, but then two copies of the P pentomino would be required to complete the construction. This completes the proof that the construction required in problem 81 is not possible.

POSTSCRIPT: As of October 1993, three solutions to problem 88 have been announced, all concluding that the required construction is impossible. Andrew Vancil, a graduate student of Professor Bennet Manvel of Colorado State University at Fort Collins, wrote a master's thesis titled "The Pentomino Fifteen Problem." Lorne Lugger of Alabama has an independent solution, and Charles Ashbacher of Cedar Rapids has run an exhaustive computer search. The latter two have joined forces and submitted a paper to the *Journal of Recreational Mathematics* about their results.

ACKNOWLEDGMENT: The author thanks Richard Guy for references [2, 12, 15], Kathy Jones [13] and Nob Yoshigahara [18] for their oral communications, and Martin Gardner [6] for his informative letter.

REFERENCES

[1] C. J. Bouwkamp. Simultaneous 4 by 5 and 4 by 10 pentomino rectangles. *J. Recreational Math.* 3 (1970): 125.

[2] T. R. Dawson and W. E. Lester. A notation for dissection problems (III). *Fairy Chess Review* 5 (1937): 46–47.

[3] H. E. Dudeney. *The Canterbury Puzzles,* 119–21. Dover, New York, 1958.

[4] R. A. Fairbairn, letter to Martin Gardner, 4 August 1967.

[5] A. A. Garcia, letter to Martin Gardner, 9 July 1971.

[6] Martin Gardner, letter to Andy Liu, 8 May 1982.

[7] Martin Gardner. *The Scientific American Book of Mathematical Puzzles and Diversions,* 124–40. Fireside Books, New York, 1959.

[8] Martin Gardner. *New Mathematical Diversions from Scientific American,* 150–61. Fireside Books, New York, 1966.

[9] Martin Gardner. *Mathematical Magic Show,* 172–87. Mathematical Association of America, Washington, D.C., 1990.

[10] S. W. Golomb. Checkerboards and polyominoes. *Amer. Math. Monthly* 61 (1954): 675–682.

[11] S. W. Golomb. *Polyominoes,* 165–67. Charles Scribners' Sons, New York, 1965.

[12] C. B. Haselgrove and J. Haselgrove. A computer program for pentominoes. *Eureka* 23 (1960): 16–18.

[13] Kathy Jones, oral communication, 30 July 1986.

[14] Andy Liu. Pentomino problems. *J. Recreational Math.* 15 (1982): 8–13.

[15] J.C.P. Miller. Pentominoes. *Eureka* 23 (1960): 13–16.

[16] J.M.M. Verbakel. The F-pentacube problem. *J. Recreational Math.* 5 (1972): 20–21.

[17] M. R. Walsh, letter to Martin Gardner, 13 March 1975.

[18] Nob Yoshigahara, oral communication, 30 July 1986.

D

Klarner's Konstant and the Enumeration of N-Ominoes

In his 1966 doctoral thesis, David A. Klarner proved the existence of a constant K such that the total number of n-ominoes, $P(n)$, grows like K^n as n goes to infinity. (Specifically, he showed that the nth root of $P(n)$ has a limit as n goes to infinity, and we call this limit "Klarner's Konstant" K.) Klarner was able to show that K is at least 3.72, and Murray Eden of M.I.T. had already shown that K is at most 6.75. Both of these results have undergone successive improvements. The current lower bound (Klarner and Wade Satterfield) is 3.9^+, and the current upper bound (Klarner and Ronald Rivest) is 4.649551. The truth lies somewhere in between, perhaps near 4.2.

The exact value of $P(n)$ has now been determined up to $n = 24$. After the work of R. C. Read and Tom Parkin (up to $n = 10$) mentioned in chapter 6, major advances were made by W. F. Lunnon (up to $n = 15$ for $P(n)$ and $P^*(n)$) and by D. H. Redelmeier (up to $n = 24$ for $P(n)$), as summarized in the table D.1. Here $P(n)$ is the *total* number of n-ominoes, and $P^*(n)$ is the number of n-ominoes without holes.

Table D.1

n	$P(n)$	$P^*(n)$	n	$P(n)$	$P^*(n)$	n	$P(n)$
1	1	1	9	1,285	1,248	17	50,107,909
2	1	1	10	4,655	4,460	18	192,622,052
3	2	2	11	17,073	16,094	19	742,624,232
4	5	5	12	63,600	58,937	20	2,870,671,950
5	12	12	13	238,591	217,117	21	11,123,060,678
6	35	35	14	901,971	805,475	22	43,191,857,688
7	108	107	15	3,426,576	3,001,211	23	168,047,007,728
8	369	363	16	13,079,255		24	654,999,700,403

It is an unproved conjecture that the ratio $P(n + 1)/P(n)$ steadily increases as n increases, approaching Klarner's Konstant K from below.

Starting in late 1992, Matthew Tibbits, a high school student in Asheville, North Carolina, undertook the computer enumeration of "polycubes," the solid polyominoes introduced in chapter 6. Table D.2 shows his results up to $n = 8$, where $S_0(n)$ is the number of different shapes made of n unit cubes when we do not count mirror images as distinct, and $S(n)$ is the larger number of shapes when mirror images that cannot be brought into coincidence by rigid motions in 3-dimensional space are counted separately. The values of $S(n)$ up to $n = 12$ are previously unpublished results of W. F. Lunnon.

There is a very general combinatorial principle that as n increases, the percentage of objects with any type of symmetry decreases. It is a type of symmetry for a 3-dimensional object to be the same as its mirror image. Accordingly, the *ratio* of $S(n)$ to $S_0(n)$ can be expected to approach closer and closer to 2 as n continues to increase, although the *difference* $2S_0(n) - S(n)$ is also likely to continue to increase. (This difference is the number of solid n-ominoes that *are* the same as their mirror images, within the limitations of rigid motions in 3-dimensional space.)

An asymptotic bound for the number of "convex" polyominoes was published by Klarner and Rivest in 1974.

TABLE D.2

n	1	2	3	4	5	6	7	8	9	10	11	12
$S_0(n)$	1	1	2	7	23	114	625	3974				
$S(n)$	1	1	2	8	29	166	1023	6922	48311	346543	2522572	18598427

REFERENCES

M. Eden. A two-dimensional growth process. *Proc. Fourth Berkeley Sympos. on Math. Statistics and Probability* 4 (1961): 223–39.

D. A. Klarner. Cell growth problems. *Canad. J. Math.* 19 (1967): 851–63.

D. A. Klarner and R. Rivest. A procedure for improving the upper bound for the number of n-ominoes. *Canad. J. Math.* 25 (1973): 585–602.

———. Asymptotic bounds for the number of convex n-ominoes. *Discrete Math.* 8 (1974): 31–40.

W. F. Lunnon. Counting polyominoes. In A. O. L. Atkin and B. J. Birch, eds., *Computers in Number Theory*, 347–72. Academic Press, London, 1971.

D. H. Redelmeier. Counting polyominoes: Yet another attack. *Discrete Math.* 36, no. 2 (1981): 191–203.

Glossary

A

Animal. A geometric figure composed of identical elements or "cells."

B

Backtrack. A systematic search procedure for favorable configurations that uses trial and error to build up a configuration one element at a time.

Base. A number that is to be raised to a certain power, which is indicated by the superscript exponent.

Binomial Coefficients. The numbers $\binom{n}{k}$ that appear as coefficients in the binomial theorem and that also equal the number of ways to select a subset of k objects from a set of n objects.

Binomial Theorem. The algebraic formula:

$$(x + y)^n = x^n + \binom{n}{1} x^{n-1}y + \binom{n}{2} x^{n-2}y^2 + \cdots + \binom{n}{n-1} xy^{n-1} + \binom{n}{n} y^n.$$

Board. A rectangular array composed of equal-sized squares on which polyominoes may be fitted.

C

Checkerboard Coloring. The assignment of two colors to a board in such a way that adjacent squares are always of different colors.

Coefficient. A multiplier, or factor, in a mathematical term.

Combinatorial Analysis. The branch of mathematics dealing with geometric or numerical patterns, arrangements, permutations, combinations, and enumerations.

Combinatorial Geometry. The branch of combinatorial analysis concerned with geometric patterns.

Combination. An assortment of objects; for example, a social security number is a combination of decimal digits. An *ordered combination* is one in which the order of the objects is significant; a social security number is an ordered combination of nine digits. In an *unordered combination,* the order in which the objects are arranged is not significant.

Cyclic Group. The symmetry group of a regular n-sided polygon consisting of the symmetry operators that rotate the figure through angles of 0 degrees, $\frac{1}{n} \cdot 360$, $\frac{2}{n} \cdot 360$, $\frac{3}{n} \cdot 360$, ... and $\frac{n-1}{n} \cdot 360$ degrees around its center (for a symmetry group of n symmetry operators).

D

Dekomino. A polyomino composed of ten squares; there are 4655 different dekominoes.

Diamond. A triangular animal composed of two equilateral triangles.

Digit. A basic symbol used in the representation of numbers. In the decimal system, there are ten different digit symbols: 0, 1, 2, 3, 4, 5, 6, 7, 8, and 9.

Dihedral Group. The symmetry group of a regular n-sided polygon containing all n rotation symmetries of the cyclic group in addition to the n reflection symmetries corresponding to reflecting the figure in any one of its n diagonals (the lines through the center of the figure terminating either in its vertices or in the midpoints of its sides).

Domino. A polyomino composed of two squares; it has only one shape.

E

Enneomino. A nonomino.

Enumeration. Counting; the determination of the number of distinct cases, patterns, or arrangements in a set.

Exponent. A numerical superscript of a number indicating its power or the number of times the number should be multiplied by itself.

F

Factorial. With a whole number n, the product of all the integers from 1 to n inclusive. It equals the number of permutations on n distinct objects and is denoted by $n!$.

G

Grid Line. A horizontal or vertical line across a board that does not intersect any square but runs between the edges of two adjoining squares.

Group. A set G of mathematical objects (such as numbers or symmetry operators), with a multiplication rule satisfying the following four requirements: (1) *Closure:* If a and b are objects in G, then $a \times b$ is a uniquely defined object in G. (2) *Associativity:* if a, b, and c are objects in G, then $a \times (b \times c) = (a \times b) \times c$. (3) *Identity element:* There is an object e in G such that $e \times a = a \times e = a$ for every a in G. This object e is called the identity element of G. (4) *Inverse element:* If a is an object in G, then there is an object a' in G such that $a \times a' = a' \times a = e$, where e is the identity element of G; a' is called the inverse of a.

Note: In general, group elements need not satisfy the *commutative law*, $a \times b = b \times a$. If this law is obeyed by all objects a and b of G, G is called a *commutative group*.

H

Heptiamond. A triangular animal composed of seven equilateral triangles; there are twenty-four different heptiamonds.

Heptomino. A polyomino composed of seven squares; there are 108 different heptominoes.

Hexagonal Animal. A pattern formed by a specific number of equal-sized regular hexagons connected by common edges. (These are also called *polyhexes*.)

Hexiamond. A triangular animal composed of six equilateral triangles; there are twelve different hexiamonds.

Hexomino. A polyomino composed of 6 squares; there are thirty-five different hexominoes.

I

Integer. A whole number. The positive integers are the numbers 1, 2, 3, 4, ...; the negative integers are -1, -2, -3, -4, ...; and the remaining integer is 0.

M

Mathematical Induction. The name given to the deductive principle that: If $P(n)$ is a statement about the positive integers n, if it can be shown that $P(1)$ is true, and if it can be shown that the truth of $P(n + 1)$ follows from the truth of $P(n)$ for all positive integers n, then $P(n)$ is true in all cases.

Monomino. A single square; a polyomino of one square.

N

Nonomino. A polyomino composed of nine squares; there are 1285 different nonominoes (also called an *enneomino*).

Number. A mathematical designator of quantity.

O

Octomino. A polyomino composed of eight squares; there are 369 different octominoes.

Order of a polyomino. The smallest number of identical copies of the polyomino which can be assembled to form a rectangle.

Ordered Combination. See Combination.

Odd-order of a polyomino. The smallest odd number of identical copies of the polyomino which can be assembled to form a rectangle.

P

Parity. The property of a number being either even or odd.

Parity Check. A test for the consistency of a mathematical statement based solely on the parity of the numbers involved.

Pentacube. A pattern composed of five equal-sized cubes connected by common faces; there are twenty-nine different pentacubes.

Pentiamond. A triangular animal composed of five equilateral triangles. There are four different pentiamonds.

Pentomino. A polyomino composed of five squares; there are twelve possible pentominoes.

Permutation. A rearrangement or renumbering of the objects in a set.

Polygon. A closed plane figure bounded by straight lines.

Polyhex. A hexagonal animal.

Polyiamond. A plane figure composed of equilateral triangles sharing common edges.

Polyomino. A pattern formed by the connection of a specified number of equal-sized squares along common edges.

Pseudo-Polyomino. A pattern formed by several equal-sized squares connected either by common edges or joined at right angles by common vertices.

Q

Quasi-Polyomino. A pattern formed by several squares that need not even be connected but that can arise as a subset of the squares in a polyomino.

R

Reflection. Flipping over a geometric figure relative to a line or plane within it.

Rep-tile. A geometric figure such that a finite number of identical copies of it can be assembled to form an enlarged scale model of it.

Rotation. Angular motion of a geometric figure relative to a point or line within it.

S

Set. Any collection of objects. Frequently, the objects in a set have some common characteristic; for example, all the dogs in China could be considered to be a set.

Solid Polyomino. A pattern formed by several equal-sized cubes connected by common faces; specifically, the figure resulting when the squares of a polyomino are replaced by cubes.

Subset. Part of a set, usually with some particular characteristic common to all its member objects. For example, the even numbers form a subset of the set of whole numbers.

Symmetry. A property of figures or configurations whereby the object looks the same after certain of its parts are interchanged.

Symmetry Group. A group composed of the symmetry operators of a particular geometric figure or mathematical configuration, where the "product" of two symmetry operators is that operator achieving in one step the same effect as that obtained by performing the two original symmetry operations one after the other. For example, the "product" of a 90-degree rotation and a 180-degree rotation is a 270-degree rotation.

Symmetry Operator. A systematic interchanging of parts of a figure in such a way that its appearance remains unchanged; for example, rotating a square 90 degrees around its center.

T

Tetromino. A polyomino composed of four squares; there are five different tetrominoes.

Tile. A geometric figure such that identical copies of it can be used to cover a specified region of the plane without gaps or overlaps.

Triangular Animal. A pattern formed by several equal-sized equilateral triangles connected by common edges.

Tromino. A polyomino composed of three squares; there are two different trominoes.

U

Unordered Combination. See Combination.

W

Whole Number. See Integer.

Bibliography for the First Edition

(Note: In this original bibliography, entries within sections were arranged chronologically, by month and year of publication.)

ARTICLES ON POLYOMINOES

Scientific American, "Mathematical Games" column by Martin Gardner

"About the Remarkable Similarity between the Icosian Game and the Tower of Hanoi." CXCVI (May, 1957), 5, 154–56.

"More about Complex Dominoes, Plus the Answers to Last Month's Puzzles." CXCVII (December, 1957), 6, 126–29.

"A Game in Which Standard Pieces Composed of Cubes Are Assembled into Larger Forms" (Soma Cubes). CXCIX (September, 1958), 3, 182–88.

"More about the Shapes That Can Be Made with Complex Dominoes." CCIII (November, 1960), 5, 186–94.

"A New Collection of Brain Teasers." CCIV (June, 1961), 6, 168.

"Some Puzzles Based on Checkerboards and Answers to Last Month's Problems." CCVII (November, 1962), 5, 151–59.

Recreational Mathematics Magazine

Golomb, Solomon W. "The General Theory of Polyominoes, Part 1, Dominoes, Pentominoes, and Checkerboards." No. 4 (August, 1961), 3–12.

——. "The General Theory of Polyominoes, Part 2, Patterns and Polyominoes." No. 5 (October, 1961), 3–12 (*see also* Notes, 13–14).

——. "The General Theory of Polyominoes, Part 3, Pentomino Exclusion by Monominoes." No. 6 (December, 1961), 3–20.

——. "The General Theory of Polyominoes, Part 4, Extensions of Polyominoes." No. 8 (April, 1962), 7–16.

Anderson, Jean H. "Polyominoes—The 'Twenty Problem.'" No. 9 (June, 1962), 25–30.

"Polyominoes—The 'Twenty Problem' and Others." No. 10 (August, 1962), 25–28.

New Scientist, "Puzzles and Paradoxes" column by T. H. O'Beirne

"Pell's Equation in Two Popular Problems." No. 258 (October 26, 1961), 260–61.

"Pentominoes and Hexiamonds." No. 259 (November 2, 1961), 316–17.

"Some Hexiamond Solutions and an Introduction to a Set of 25 Remarkable Points." No. 260 (November 9, 1961), 379–80.
"For Boys, Men and Heroes." No. 266 (December 21, 1961), 751–52.
"Some Tetrabolical Difficulties." No. 270 (January 18, 1962), 158–59.

Fairy Chess Review

Dawson, T. R., and W. E. Lester. "A Notation for Dissection Problems." III (April, 1937), 5, 46–47.
Stead, W. "Dissection." IX (December, 1954), 1, 2–4.

Various Mathematics Journals

Golomb, Solomon W. "Checkerboards and Polyominoes." *American Mathematical Monthly* LXI (December, 1954), 10, 672–82.
Reeve, J. E., and J. A. Tyrrell. "Maestro Puzzles." *Mathematical Gazette* XLV (1961), 97–99.
Read, R. C. "Contributions to the Cell Growth Problem." *Canadian Journal of Mathematics* XIV (1962), 1, 1–20.
Hajtman, Bela. "On Coverings of Generalized Checkerboards I." *Magyar Tud. Akad. Mat. Kutato Int. Köze.* VII (1962), 53–71.

BOOKS CONTAINING POLYOMINO MATERIAL

Dudeney, H. E. *The Canterbury Puzzles.* Problem 74, 119–20. New York: Dover Publications, Inc., 1958.
Gardner, Martin. *The Scientific American Book of Mathematical Puzzles and Diversions,* 124–40. New York: Simon and Schuster, 1959.
——. *The Second Scientific American Book of Mathematical Puzzles and Diversions,* 65–77. New York: Simon and Schuster, 1961.
Hunter, J. A., and Joseph S. Madachy. *Mathematical Diversions,* 77–89. New York: D. Van Nostrand Company, Inc., 1963.

MORE TECHNICAL COMBINATORIAL MATERIAL IN ARTICLES AND BOOKS

Scott, Dana S. "Programming a Combinatorial Puzzle." Unpublished study, Department of Electrical Engineering. Princeton University, Princeton, N.J., June 10, 1958.
Riordan, John. *An Introduction to Combinatorial Analysis.* New York: John Wiley, Inc., 1958.
Golomb, Solomon W. "A Mathematical Theory of Discrete Classification," 404–25. *Fourth London Symposium on Information Theory.* London: Butterworth, 1961.

Comprehensive Bibliography

PROCEEDINGS

Acharya, B. Devadas [Devadas Acharya, B.]. "Are All Polyominoes Arbitrarily Graceful?" In *Graph Theory*, 205–11. Proceedings of the First Southeast Asian Colloq. [Singapore, 1983]. Lecture Notes in Mathematics Series, no. 1073. Berlin–New York: Springer-Verlag, 1984.

Akiyama, Jin, and Mikio Kano. "Path Factors of a Graph." In *Graphs and Applications*, 1–21. [Boulder, Colo., 1982]. New York: Wiley, 1985.

Akiyama, Jin, Mikio Kano, and Mari-Jo Ruiz. "Tiling Finite Figures Consisting of Regular Polygons." In *Graph Theory with Applications to Algorithms and Computer Science*, 1–13. Proceedings of the 5th International Conference [Kalamazoo, Mich., 1984]. New York: Wiley, 1985.

Boros, E. "On the Number of Subdivisions of the Unit Square." In *Finite and Infinite Sets*. Proceedings of the 6th Hungarian Combin. Colloq. [Eger, Hungary, 1981]. *Colloq. Math. Soc. Janos Bolyai* 2, no. 37 (1984): 893–98.

Chinn, Phyllis, and Dan Munton. "On the Edge-connectivity and Separation Sequences of Rectangular Grid Graphs." Nineteenth Southeastern Conference on Combinatorics, Graph Theory, and Computing [Baton Rouge, La., 1988]. *Congressus Numerantium* 65 (1988).

Delest, Marie-Pierre. "Enumeration of Polyominoes Using Macsyma." Algebraic and Computing Treatment of Noncommutative Power Series [Lille, 1988]. *Theoretical Computer Science* 79, no. 1, Part A (1991): 209–26.

Delest, Marie-Pierre. "Polyominoes and Animals: Some Recent Results." Paper presented at Mathematical Chemistry and Computation [Dubrovnik, 1990]. *Journal of Mathematical Chemistry* 8, nos. 1–3 (1991): 3–18.

Delest, Marie-Pierre, and J. M. Fedou. "Counting Polyominoes Using Attribute Grammars." Workshop: Attribute Grammars and their Applications [Paris, 1990]. Lecture Notes in Computer Science Series, no. 461. Berlin–New York: Springer-Verlag, 1990.

Delest, Marie-Pierre, and Gerard Viennot. "Algebraic Languages and Polyominoes Enumeration." In *Automata, Languages and Programming*. 173–81. Proceedings of the 10th Colloq. [Barcelona, 1983]. Lecture Notes in Computer Science Series, no. 154. Berlin–New York: Springer-Verlag, 1983.

Eden, M. "A Two-Dimensional Growth Process," 223–39. Proceedings of the Fourth Berkeley Symposium on Mathematical Statistics and Probability, vol. 4 [Berkeley, Calif., 1961].

Goebel, F. "Geometrical Packing and Covering Problems." In *Packing and Covering in Combinatorics*, Study Week [Amsterdam, 1978]. *Math. Cent. Tracts* 106 (1979): 179–99.

Golomb, Solomon W. "A Mathematical Theory of Discrete Classification." In *Fourth London Symposium on Information Theory*, 404–25. London: Butterworth, 1961.

Gutman, Ivan, and Jerry R. Dias. "The Excised Internal Structure of Hexagonal Systems." In *Contemporary Methods in Graph Theory*. In honor of Prof. K. Wagner. Conference article (1990): 249–59.

Harary, Frank. "An Achievement Game on a Toroidal Board." In *Graph Theory*, 55–59. [Lagow, Poland, 1981]. Lecture Notes in Mathematics Series, no. 1018. Berlin–New York: Springer-Verlag, 1983.

Harborth, Heiko, and Sabine Lohmann. "Mosaic Numbers of Fibonacci Trees." In *Applications of Fibonacci Numbers* 3: 133–38. Proceedings of the 3d International Conference [Pisa, 1988]. Dordrecht: Kluwer Academic Publishers, 1990.

Harborth, Heiko, and Hartmut Weiss. "Minimum Sets of Partial Polyominoes." Sixteenth Australasian Conference [Palmerston North, 1990]. *Australasian Journal of Combinatorics* 4 (1991).

Holladay, Kenneth. "On 3-irreducible Animals." In *Combinatorics, Graph Theory and Computing*, vol. 2. Proceedings of the 10th Southeast Conference [Boca Raton, Fla., 1979]. *Congressus Numerantium* 24 (1979): 511–21.

Huber-Stockar, Emil. "L'echiquier du diable." In *Comptes-Rendus du Deuxième Congrès International de Récréation Mathématique*, 64–68. [Paris, 1937]. Brussels: Librairie du Sphinx, 1937.

———. "Patience de l'echiquier." In *Comptes-Rendus du Premier Congrès International de Récréation Mathématique*, 93–94. [Brussels, 1935]. Brussels: Librairie du Sphinx, 1935.

John, Peter Eugen, Horst Sachs, and H. Zernitz. "Counting Perfect Matchings in Polyominoes with an Application to the Dimer Problem." International Conference on Combinatorial Analysis and Its Applications [Pokrzywna, 1985]. Polska Akademia Nauk. Instytut Matematyczny. *Zastosowania Matematyki* 19, nos. 3–4 (1987): 465–77.

Klarner, David A. "Methods for the General Cell Growth Problem." In *Combinatorial Theory and Its Applications. Colloq. Math. Soc. János Bolyai*. Balatonfüred. (1969): 705–20.

Kriz, Igor. "N-tilability of Acyclic Polyominoes." Proceedings of the 12th Winter School on Abstract Analysis [Srni, 1984]. *Rendiconti del Circolo Matematico di Palermo*. Series 2, suppl. no. 6 (1984): 189–200.

Kuperberg, W. "On Packing the Plane with Congruent Copies of a Convex Body." In *Intuitive Geometry*. Proceedings of Pap. Int. Conf. [Siofok, Hungary, 1985]. *Colloq. Math. Soc. János Bolyai* 48 (1987): 317–29.

Comprehensive
Bibliography

Mazumder, P., and J. Tartar. "Planar Topologies for Quadtree Representation." Fourteenth Manitoba Conference on Numerical Mathematics and Computing [Winnipeg, 1984]. *Congressus Numerantium* 46 (1985).

Reid, K. B. "A Polyominoe Labeling Problem." *Proceedings of the Louisiana Conference on Combinatorics, Graph Theory and Computing.* Louisiana State University, Baton Rouge (1970): 227–48.

Robinson, Peter J. "Fault-free Rectangles Tiled with Rectangular Polyominoes." In *Combinatorial Mathematics,* Proceedings of the Ninth Australian Conference [Brisbane, 1981]. Lecture Notes in Mathematics Series, no. 952. Berlin–New York: Springer-Verlag, 1982.

Tilley, R. C., R. G. Stanton, and D. D. Cowan. "The Cell Growth Problem for Filaments." *Proceedings of the Louisiana Conference on Combinatorics, Graph Theory and Computing.* Louisiana State University, Baton Rouge (1970): 310–39.

Tosic, Ratko, Rade Doroslovacki, and Ivan Stojmenovic. "Generating and Counting Square Systems." In *Graph Theory,* 127–36. Proceedings of the 8th Yugosl. Semin. [Novi Sad, Yugoslavia, 1987]. Novi Sad: Univ. Novi Sad, 1989.

Viennot, Gerard. "Enumerative Combinatorics and Algebraic Languages." In *Fundamentals of Computation Theory,* 450–64. Proceedings of the 5th International Conference [Cottbus, Germany, 1985]. Lecture Notes in Computer Science Series, no. 199. Berlin–New York: Springer-Verlag, 1985.

BOOKS

Association of Teachers of Mathematics. *Mathematical Reflections: Contributions to Mathematical Thought and Teaching, written in memory of A. G. Sillitto.* Cambridge: Cambridge University Press, 1970.

Bent, S. W. *Stable Transversals and Stochastic Functions in Polyominoes.* Stanford Technical Reports, Report 520, 1983.

Berlekamp, E. R., J. H. Conway, and R. K. Guy. *Winning Ways for Your Mathematical Plays,* 786–88. New York: Academic Press, 1982.

Black, Max. *Critical Thinking. An Introduction to Logic and Scientific Method.* New York: Prentice-Hall, 1946, pp. 142, 394; 2d ed. 1952, pp. 157, 433.

Boltyanski, Vladimir, and Alexander Soifer. *Geometric Etudes in Combinatorial Mathematics.* Colorado Springs: Center for Excellence in Mathematical Education, 1991.

Bouwkamp, C. J. *Catalogue of Solutions of the Rectangular 3 × 4 × 5 Solid Pentomino Problem.* Eindhoven, The Netherlands: Department of Mathematics, Technological University, Eindhoven, July 1967. Reprinted 1981.

———. *Packing the Steps with Solid Pentominoes.* Technische Hogeschool Report 79. WSK-01. Eindhoven, The Netherlands: Department of Mathematics, Technological University, Eindhoven, 1979.

Clarke, Arthur C. *Ascent to Orbit*. New York: Wiley, 1984.

———. *Imperial Earth*. New York: Harcourt Brace Jovanovich, 1976.

Collins, A. Frederick. *The Book of Puzzles*, 131–34. New York: D. Appleton & Co., 1927.

Croft, Hallard T., Kenneth J. Falconer, and Richard K. Guy. *Unsolved Problems in Geometry*. New York: Springer-Verlag, 1991.

Croft, Hallard T., Kenneth J. Falconer, and Richard K. Guy. *Unsolved Problems in Intuitive Mathematics*. New York: Springer-Verlag, 1981.

Bruijn, N. G. de, and D. A. Klarner. "A Finite Basis Theorem for Packing Boxes with Bricks." In *Philips Research Reports 30, Papers Dedicated to C. J. Bouwkamp* (1975): 337–43.

Dudeney, Henry Ernest. "The Broken Chessboard." Problem 74 in *The Canterbury Puzzles and Other Curious Problems,* 119–21, 220–21. New York: Dover Publications, 1958.

———. "The Chinese Chessboard." Problem 293 in *Amusements in Mathematics*, 87, 213–14. New York: Dover Publications, 1970.

Farhi, Sivy. *Pentacubes*. New Zealand: Booklet published by the author, 1981.

———. *Pentominoes*. New Zealand: Booklet published by the author, 1981.

———. *Soma Cubes*. New Zealand: Booklet published by the author, 1979.

———. *Soma World: The Complete Soma Cube*. New Zealand: Booklet published by the author, 1982.

Gardner, Martin. "Animal TTT." "Sex Among the Polyomans." In *Riddles of the Sphinx and Other Mathematical Puzzle Tales*. Washington, D.C.: Mathematical Association of America, 1987.

———. "Cutting Shapes into Congruent Parts." In *Penrose Tiles to Trapdoor Ciphers*. New York: W. H. Freeman, 1989.

———. "Dominoes." In *Mathematical Circus*. New York: Knopf, 1979.

———. "The Eight Queens and Other Chessboard Diversions." "Rep-Tiles: Replicating Figures in the Plane." In *The Unexpected Hanging, and Other Mathematical Diversions*. New York: Simon and Schuster, 1969.

———. "Hypercubes." In *Mathematical Carnival*. New York: Knopf, 1975.

———. "Polycubes." "The Tour of the Arrows and Other Problems." "Reverse the Fish and Other Problems." "Cram, Bynum and Quadraphage." In *Knotted Doughnuts and Other Mathematical Entertainments*. New York: W. H. Freeman, 1986.

———. "Polyhexes and Polyaboloes." "Polyominoes and Rectification." "Polyomino Four-color Problem." In *Mathematical Magic Show*. New York: Knopf, 1977.

———. "Polyiamonds." "Op Art." In *Sixth Book of Mathematical Games from Scientific American*. New York: W. H. Freeman, 1971. Rev. ed., Chicago: University of Chicago Press, 1983.

Comprehensive
Bibliography

————. "Polyominoes." In *Mathematical Puzzles and Diversions*. New York: Simon and Schuster, 1959.

————. "Polyominoes and Fault-free Rectangles." "Problem: Two Pentomino Posers." In *New Mathematical Diversions from Scientific American*. New York: Simon and Schuster, 1966.

————. *The Scientific American Book of Mathematical Puzzles and Diversions*, 124–40. New York: Simon and Schuster, 1959.

————. "The Soma Cube." In *The Second Scientific American Book of Mathematical Puzzles and Diversions*. New York: Simon and Schuster, 1961.

————. "Tiling with Polyominoes, Polyiamonds, and Polyhexes." "Block Packing." In *Time Travel and Other Mathematical Bewilderments*. New York: W. H. Freeman, 1988.

Gardner, Martin, ed. *The Mathematical Puzzles of Sam Loyd*. New York: Dover Publications, 1959.

————. *More Mathematical Puzzles of Sam Loyd*. New York: Dover Publications, 1960.

Golomb, Solomon W. *Polyominoes*. New York: Scribners, 1965.

Graham, Ronald L., Donald E. Knuth, and Oren Patashnik. *Concrete Mathematics*. Reading, MA: Addison-Wesley, 1989. Second edition, 1994.

Grünbaum, Branko, and G. C. Shephard. "Some Problems on Plane Tilings," in *The Mathematical Gardner*, edited by David A. Klarner, 167–96. Boston: Prindle, Weber and Schmidt, 1981.

Grünbaum, Branko, and G. C. Shephard. *Tilings and Patterns*. New York: W. H. Freeman, 1987.

Haemers, W. H. *Eigenvalue Techniques in Design and Graph Theory*. Amsterdam: Mathematisch Centrum. V, Mathematical Centre Tracts, 1980.

Hall, A. Neely. "The Square Puzzle." In *Carpentry and Mechanics for Boys*, 20–21. Boston: Lothrop, Lee and Shepard, n.d. [1918].

Harary, Frank. "How Many Animals Are There?" In *Graph Theory and Theoretical Physics*. New York: Academic Press, 1967.

Honsberger, Ross. "Box-packing Problems." In *Mathematical Gems II*, 58–89. Washington, D.C.: Mathematical Association of America, 1976.

Hunter, J. A., and Joseph S. Madachy. "Fun with Shapes." In *Mathematical Diversions*, 77–89. New York: D. Van Nostrand Company, Inc., 1963.

Klarner, David A. "My Life Among the Polyominoes." In *The Mathematical Gardner*, edited by David A. Klarner. Boston: Prindle, Weber and Schmidt, 1981.

Larsen, Mogens Esrom. *There Is No Gap in a Diamond*. Pamphlet 19, København Universitets Matematisk Institut, October 1986.

Lines, Malcolm E. *Think of a Number: Ideas, Concepts and Problems Which Challenge the Mind and Baffle the Experts*. New York: Adam Hilger, Ltd., 1990.

Loyd, Sam. "A Battle Royal." "A Tailor's Problem." In *Cyclopedia of 5,000 Puzzles, Tricks and Conundrums*. New York: Lamb Publishing Co., 1914.

Loyd, Sam, Jr. "Wrangling Heirs." In *Sam Loyd and His Puzzles,* 35, 96. New York: Barse & Co., 1928.

Lunnon, W. F. "Counting Hexagonal and Triangular Polyominoes." In *Graph Theory and Computing,* edited by R. C. Read, 87–100. New York: Academic Press, 1972.

Lunnon, W. F. "Counting Polyominoes." In *Computers in Number Theory,* edited by A.O.L. Atkin and B. J. Birch, 347–72. London: Academic Press, 1971.

Lunnon, W. F. "Symmetry of Cubical and General Polyominoes." In *Graph Theory and Computing,* edited by R. C. Read, 101–8. New York: Academic Press, 1972.

Martin, George E. *Polyominoes. A Guide to Puzzles and Problems in Tiling.* Washington, D.C.: Mathematical Association of America, 1991.

Meeus, Jean, and Pieter J. Torbijn. *Polycubes.* France: CEDIC, 1977.

Page, Donald D. *Problem Solving with Ominoes. For Grades 9–12.* Booklet. Menlo Park, Calif.: Dale Seymour Publications, 1987.

Percus, Jerome Kenneth. *Combinatorial Methods.* New York: Springer-Verlag, 1971.

Raven, Robert S. (proposer); Walter P. Targoff (solver). "Problem 85—Deleted Checkerboard." In L. A. Graham, *Ingenious Mathematical Problems and Methods,* 52, 227. New York: Dover Publications, 1959.

Riordan, John. *An Introduction to Combinatorial Analysis.* New York: Wiley, 1958.

Scott, Dana S. *Programming a Combinatorial Puzzle.* Technical Report No. 1. Unpublished Study, June 10, 1958. Princeton, N.J.: Department of Electrical Engineering, Princeton University.

Slocum, Jerry. *Compendium of Checkerboard Puzzles.* Booklet published by the author, 1983. Revised (with J. Haubrich), 1993.

———. *Puzzles Old and New.* Seattle: Washington University Press, 1986.

Slothouber, Jan, and William Graatsma. *Cubics.* Deventer, Holland: Octopus Press, 1970.

Steinhaus, H. "Mikusiński's Cube." In *Mathematical Snapshots,* 179–81. New York: Oxford University Press, 1960.

Wilson, Marguerite. *Soma Puzzle Solutions.* Mountain View, Calif.: Creative Publications, 1973.

Yoshigahara, Nob. "Puzzlology Polyominoes." In *Puzzlart,* 70–83, 94–95. Tokyo: Published by the author, 1992.

JOURNAL ARTICLES

Ahrens, J. H. "Paving the Chessboard." *Journal of Combinatorial Theory.* Series A, 31, no. 3 (November 1981): 277–78.

Anderson, Jean H. "Spirals, Checkerboards, Polyominoes, and the Fibonacci Sequence." *Fibonacci Quarterly* 8 (1970): 90–95.

———. "Polyominoes—The 'Twenty Problem.'" *Recreational Mathematics Magazine,* no. 9 (June 1962): 25–30.

————. "Polyominoes—The 'Twenty Problem' and Others." *Recreational Mathematics Magazine*, no. 10 (August 1962): 25–28.

Arisawa, M. "Pentomino Doublets. Problem 470." *Journal of Recreational Mathematics* 9, no. 1 (1975–76): 26–27. Solution by R. I. Hess, *JRM* 10(2): 143–45.

Barnes, Frank W. "Every Hexomino Tiles the Plane." *Journal of Combinatorics, Information & System Sciences* 8, no. 2 (1983): 113–15.

————. "How Many $1 \times 2 \times 4$ Bricks Can You Get into an Odd Box?" *Discrete Mathematics*. To appear (ca. 1995).

Barnes, Frank W., and James B. Shearer. "Barring Rectangles from the Plane." *Journal of Combinatorial Theory*. Series A, 33, no. 1 (1982): 9–29.

Barwell, Brian R. "Polysticks." *Journal of Recreational Mathematics* 22, no. 3 (1990): 165–75.

Beauquier, Daniele [Girault-Beauquier, Daniele]. "An Undecidable Problem about Rational Sets and Contour Words of Polyominoes." *Information Processing Letters* 37, no. 5 (1991): 257–63.

Beauquier, Daniele, and M. Nivat. "On Translating One Polyomino to Tile the Plane." *Discrete & Computational Geometry* 6, no. 6 (1991): 575–92.

Bender, Edward A. "Convex n-Ominoes." *Discrete Mathematics* 8, no. 3 (1974): 219–26.

Bender, Edward A., L. Bruce Richmond, and S. G. Williamson. "Central and Local Limit Theorems Applied to Asymptotic Enumeration. III. Matrix Recursions." *Journal of Combinatorial Theory*. Series A, 35 (1983): 263–78.

Benjamin, H. D. Untitled. *Problemist Fairy Chess Supplement* (December 1934). (Later known as *Fairy Chess Review*).

Bent, Samuel W. "Stable Transversals and Stochastic Functions in Polyominoes." Consilium Instituti Mathematici Academiae Scientiarum Hungaricae. *Studia Scientiarum Mathematicarum* 23, nos. 1–2 (1988): 105–14.

Berge, Claude, C. C. Chen, Vashek Chvátal, and C. S. Seow. "Combinatorial Properties of Polyominoes." *Combinatorica* 1, no. 3 (1981): 217–24.

Berger, Robert. "The Undecidability of the Domino Problem." *Memoirs of the American Mathematical Society*, no. 66 (1966): 1–72.

Bestelmeier, G. H. "Das Zakken und Hakkenspiel" (Item 61) and "Das Mathematisch Kreuz" (Item 274). *Magazin von Verschiedenen Kunst—und andern nützlichen Sachen*. Nürnberg. (1801). Reprinted by Edition Olms, Zurich, 1979.

Bird, David. "The Known World of Octiamonds." *Journal of Recreational Mathematics* 8 (1975–76): 300–301.

Bitner, James. "Tiling 5×12 Rectangles with Y-Pentominoes." *Journal of Recreational Mathematics* 7, no. 4 (1974): 276–78.

Bono, Edward de. "De Bono's L-Game." *Games and Puzzles Journal* (November 1974, February 1975): 4–6, 36 (British monthly).

Bousquet-Melou, Mireille. "Une Bijection Entre les Polyominos

Convexes Diriges et les Mots de Dyck Bilateres" (A Bijection Between Convex Directed Polyominoes and Bilateral Dyck Words). *RAIRO Informatique Theorique et Applications* 26, no. 3 (1992): 205–19.

———. "Convex Polyominoes and Algebraic Languages." *Journal of Physics*. Series A. Mathematical and General 25, no. 7 (1992): 1935–44.

———. "Convex Polyominoes and Heaps of Segments." *Journal of Physics*. Series A, Mathematical and General 25, no. 7 (1992): 1925–34.

Bousquet-Melou, Mireille, and Xavier Gerard Viennot. "Empilements de Segments et q-énumération de Polyominos Convexes Dirigés" (Stacks of Segments and q-enumeration of Directed Convex Polyominoes). *Journal of Combinatorial Theory*. Series A, 60, no. 2 (1992): 196–224.

Bouwkamp, C. J. "Catalogue of Solutions of the Rectangular 2 × 5 × 6 Solid Pentomino Problem." Koninklijke Nederlandse Akademie van Wetenschappen. *Indagationes Mathematicae* 40, no. 2 (1978): 177–86.

———. "Packing a Rectangular Box with the Twelve Solid Pentominoes." *Journal of Combinatorial Theory* 7 (1969): 278–80.

———. "Simultaneous 4 × 5 and 4 × 10 Pentomino Rectangles." *Journal of Recreational Mathematics* 3, no. 2 (April 1970): 125.

Bouwkamp, C. J., and David A. Klarner. "Packing a Box with Y-Pentacubes." *Journal of Recreational Mathematics* 3, no. 1 (January 1970): 10–26.

Brak, R., and A. J. Guttmann. "Exact Solution of the Staircase and Row-Convex Polygon Perimeter and Area Generating Function." *Journal of Physics*. Series A, Mathematical and General 23, no. 20 (1990): 4581–88.

Brualdi, Richard A., and Thomas H. Foregger. "Packing Boxes with Harmonic Bricks." *Journal of Combinatorial Theory*. Series B, 17, no. 2 (1974): 81–114.

Bruijn, Nicholas G. de. "Filling Boxes with Bricks." *American Mathematical Monthly* 76 (1969): 37–40.

———. "Programmeren van de Pentomino Puzzle." *Euclides* 47, no. 3 (1971–72): 90–104.

Chu, I-Ping, and Richard Johnsonbaugh. "Tiling Boards with Trominoes." *Journal of Recreational Mathematics* 18, no. 3 (1985–86): 188–93.

———. "Tiling Deficient Boards with Trominoes." *Mathematics Magazine* 59, no. 1 (1986): 34–40.

Clarke, Andrew L. "Isoperimetrical Polyominoes." *Journal of Recreational Mathematics* 13, no. 1 (1980–81): 18–25.

———. "A Pentomino Conjecture. Problem 600." *Journal of Recreational Mathematics* 10, no. 1 (1977–78): 54. Solution by M. Beeler, *JRM* 12(1): 63–64.

Coll, Pablo E. "A Pentomino Problem. Problem 1277." *Journal of Recreational Mathematics* 16, no. 1 (1983–84): 62–63. Solution by Friend H. Kierstead, Jr., *JRM* 17(1): 75–77; solution by Jean

Comprehensive Bibliography

Meeus, *JRM* 18(1): 48–49; solution by Stan Vejmola, *JRM* 18(1): 49.

———. "Pentomino Problem III. Problem 1347." *Journal of Recreational Mathematics* 16, no. 4 (1983–84): 302. Solutions, *JRM* 17(4): 310–11.

Coll, Pablo E., and George P. Jelliss. "A Pentomino Problem." *Journal of Recreational Mathematics* 22, no. 1 (Spring 1990): 69.

Coll, Pablo E., and Jean Meeus. "A Pentomino Problem." *Journal of Recreational Mathematics* 21, no. 1 (Spring 1989): 69.

Conway, J. H., and J. C. Lagarias. "Tiling with Polyominoes and Combinatorial Group Theory." *Journal of Combinatorial Theory*. Series A, 53, no. 2 (1990): 183–208.

Dahlke, Karl A. "Erratum: 'A Heptomino of Order 76.'" *Journal of Combinatorial Theory*. Series A, 52, no. 2 (1989): 321.

———. "A Heptomino of Order 76." *Journal of Combinatorial Theory*. Series A, 51, no. 1 (1989): 127–28.

———. "The Y-hexomino Has Order 92." *Journal of Combinatorial Theory*. Series A, 51, no. 1 (1989): 125–26.

Davis, Roy O. "Note 3151: Replicating Boots." *Mathematical Gazette* 50, no. 372 (May 1966): 175.

Dawson, T. R., and W. E. Lester. "A Notation for Dissection Problems." *Fairy Chess Review* 3, no. 5 (April 1937): 46–47.

Dekkers, A. J., and A.J.W. Duijvestijn. "Solving a Chessboard Puzzle with the Pascal." *Philips Technical Review* 24 (1962–63): 157–63.

Delest, Marie-Pierre. "Generating Functions for Column-Convex Polyominoes." *Journal of Combinatorial Theory*. Series A, 48, no. 1 (1988): 12–31.

Delest, Marie-Pierre, D. Gouyou-Beauchamps, and B. Vauquelin. "Enumeration of Parallelogram Polyominoes with Given Bond and Site Perimeter." *Graphs and Combinatorics* 3, no. 4 (1987): 325–39 (Asian journal).

Delest, Marie-Pierre, and Xavier Gerard Viennot. "Algebraic Languages and Polyominoes Enumeration." *Theoretical Computer Science* 34, nos. 1–2 (1984): 169–206.

Denniss, J. "A Replication Problem." *Mathematics Teaching*, no. 88 (September 1979): 8–10.

Doroslovacki, Rade, Ivan Stojmenovic, and Ratko Tosic. "Generating and Counting Triangular Systems." *Nordisk Tidskrift for Informationsbehandling* [BIT] 27, no. 1 (1987): 18–24.

Dudeney, Henry Ernest. "Make a Chessboard. Problem 517." *Weekly Dispatch* 4 & 18 (October 1903): 10, 10.

Elliott, C. S. "Squaragons." *Journal of Recreational Mathematics* 13, no. 2 (1980–81): 102–7.

Ferrell, P. C. "A Pentomino Unit." *Mathematics Teacher* 70, no. 6 (September 1977): 523–27.

Feser, Fr. Victor. "Pentomino Farms." *Journal of Recreational Mathematics* 1, no. 1 (January 1968): 55–61.

Fischer, Michael E. "Statistical Mechanics of Dimers on a Plane Lattice." *Physical Review*. Series 2, 124 (1961): 1664–72.

Fletcher, John G. "A Program to Solve the Pentomino Problem by the Recursive Use of Macros." *Communications of the Association for Computing Machinery* 8, no. 10 (October 1965): 621–23.

Fontaine, Anne, and George E. Martin. "An Enneamorphic Prototile." *Journal of Combinatorial Theory*. Series A, 37, no. 2 (1984): 195–96.

————. "Polymorphic Polyominoes." *Mathematics Magazine* 57, no. 5 (November 1984): 275–83.

————. "Polymorphic Prototiles." *Journal of Combinatorial Theory*. Series A, 34 (1983): 119–21.

————. "Tetramorphic & Pentamorphic Prototiles." *Journal of Combinatorial Theory*. Series A, 34, (1983): 115–18.

Franzblau, D. S. "Performance Guarantees on a Sweep-line Heuristic for Covering Rectilinear Polygons with Rectangles." *SIAM Journal on Discrete Mathematics* 2, no. 3 (1989): 307–21.

Freeburn, R. "A New Polygon Puzzle." *Mathematics Teaching*, no. 80 (September 1977): 57–59.

————. "A New Polypolyhedron Puzzle: Some Solutions." *Mathematics Teaching*, no. 81 (December 1977): 30–31.

————. "Polypolygons." *Mathematics Teaching*, no. 74 (March 1976): 36.

French, R. J. Untitled. *Fairy Chess Review* 4, no. 3 (December 1939): 43; 4, no. 6 (June 1940): 91.

Gardner, Martin. "About the Remarkable Similarity Between the Icosian Game and the Tower of Hanoi." *Scientific American* 196, no. 5 (May 1957): Mathematical Games Column, 154–56.

————. "Dr. Matrix Returns, Now in the Guise of a Neo-Freudian Psychonumeranalyst." *Scientific American* 214, no. 1 (January 1966): Mathematical Games Column, 115.

————. "The Fantastic Combinations of John Conway's New Solitaire Game 'Life.'" *Scientific American* 223, no. 4 (October 1970): Mathematical Games Column, 120–23.

————. "A Game in Which Standard Pieces Composed of Cubes are Assembled into Larger Forms (Soma Cubes)." *Scientific American* 199, no. 3 (September 1958): Mathematical Games Column, 182–92.

————. "'Look-see' Diagrams That Offer Visual Proof of Complex Algebraic Formulas." *Scientific American* 229, no. 4 (October 1973): Mathematical Games Column, 114–18.

————. "Magic Show." *Scientific American* (June 1967): 146–59.

————. "Magic Stars, Graphs and Polyhedrons, Etc." *Scientific American* 213, no. 6 (December 1965): Mathematical Games Column, 100–105.

————. "More About Complex Dominoes, Plus the Answers to Last Month's Puzzles." *Scientific American* 197, no. 6 (December 1957): Mathematical Games Column, 126–29.

Journal Articles

————. "More about the Shapes That Can Be Made with Complex Dominoes." *Scientific American* 203, no. 5 (November 1960): Mathematical Games Column, 186–94.

————. "More About Tiling the Plane: The Possibilities of Polyominoes, Polyiamonds and Polyhexes." *Scientific American* 233, no. 2 (August 1975): Mathematical Games Column, 112–15.

————. "The Multiple Charms of Pascal's Triangle." *Scientific American* 215, no. 6 (December 1966): Mathematical Games Column, 128–32.

————. "The Mutilated Chessboard." *Scientific American* 196, no. 2 (February 1957), Mathematical Games Column, 152–56.

————. "A New Collection of Brain Teasers." *Scientific American* 204, no. 6 (June 1961): Mathematical Games Column, 168.

————. "A New Miscellany of Problems, and Encores for Race Track, Sim, Chomp and Elevators." *Scientific American* 228, no. 5 (May 1973): Mathematical Games Column, 102–7.

————. "Pentominoes and Polyominoes: Five Games and a Sampling of Problems." *Scientific American* 213, no. 4 (October 1965): Mathematical Games Column, 96–104.

————. "Pleasurable Problems with Polycubes, and the Winning Strategy for 'Slither.'" *Scientific American* 227, no. 3 (September 1972): Mathematical Games Column, 176–82.

————. "The Polyhex and the Polyabolo, Polygonal Jigsaw Puzzle Pieces." *Scientific American* 216, no. 6 (June 1967): Mathematical Games Column, 124–32.

————. "On Polyiamonds: Shapes That Are Made out of Equilateral Triangles." *Scientific American* 211, no. 6 (December 1964): Mathematical Games Column, 124–30.

————. "On the Relation Between Mathematics and the Ordered Patterns of Op Art." *Scientific American* 213, no. 1 (July 1965): Mathematical Games Column, 100–104.

————. "On 'Rep-tiles,' Polygons That Can Make Larger and Smaller Copies of Themselves." *Scientific American* 208 (1963): Mathematical Games Column, 154–64.

————. "Some Puzzles Based on Checkerboards and Answers to Last Month's Problems." *Scientific American* 207, no. 5 (November 1962): Mathematical Games Column, 151–59.

————. "Tiling the Bent Tromino with n Congruent Shapes." *Journal of Recreational Mathematics* 22, no. 3 (1990): 185–91.

Goebel, F. "The L-shape Dissection Problem. Problem 1771." *Journal of Recreational Mathematics* 22, no. 1 (1990): 64–65. Editorial comment, *JRM* 23, no. 1 (1991): 69–70; comments and partial solution by M. Beeler, *JRM* 24(1): 64–69.

Goebel, F., and A. A. Jagers. "Generalized Coverings with Polyominoes." *Journal of Recreational Mathematics* 9, no. 4 (1977): 252–57.

Golomb, Solomon W. "Checkerboards and Polyominoes." *American Mathematical Monthly* 61, no. 10 (December 1954): 675–82.

————. "The General Theory of Polyominoes, Part 1, Dominoes,

Pentominoes, and Checkerboards." *Recreational Mathematics Magazine*, no. 4 (August 1961): 3–12.

———. "The General Theory of Polyominoes, Part 2, Patterns and Polyominoes." *Recreational Mathematics Magazine*, no. 5 (October 1961): 3–12.

———. "The General Theory of Polyominoes, Part 3, Pentomino Exclusion by Monominoes." *Recreational Mathematics Magazine*, no. 6 (December 1961): 3–20.

———. "The General Theory of Polyominoes, Part 4, Extensions of Polyominoes." *Recreational Mathematics Magazine*, no. 8 (April 1962): 7–16.

———. "Hexed!" *Discover* 12, no. 12 (March 1991): Brain Bogglers, 84, 88.

———. "Polyominoes Which Tile Rectangles." *Journal of Combinatorial Theory*. Series A 51, no. 1 (1989): 117–24.

———. "Problem E1543." *American Mathematical Monthly* 69, no. 9 (November 1962): 920. Solution by D. A. Klarner, *AMM* 70, no. 7 (September 1963): 760–61.

———. "Problem 252." *Pi Mu Epsilon Journal* 5, no. 4 (Spring 1971): 203. Solutions by proposer and by Catherine Yee, *PMEJ* 5, no. 6 (Spring 1972): 300–301.

———. "Replicating Figures in the Plane." *Mathematical Gazette* 48, no. 366 (December 1964): 403–12.

———. "Tiling with Hexagons." *Johns Hopkins Magazine* 42 (October 1990): Golomb's Gambits, 11, 58.

———. "Tiling with Polyominoes." *Journal of Combinatorial Theory* 1 (1966): 280–96.

———. "Tiling with Sets of Polyominoes." *Journal of Combinatorial Theory* 9 (1970): 60–71.

Golomb, Solomon W., and Herbert Taylor. "Regions Surrounded by Polyominoes." *Journal of Recreational Mathematics* 18, no. 3 (1985–86): 214–17.

Golomb, Solomon W., and Lloyd R. Welch. "Perfect Codes in the Lee Metric and the Packing of Polyominoes." *SIAM Journal on Applied Mathematics* 18 (1970): 302–17.

Gordon, Leonard J. "Broken Chessboards with Unique Solutions." *Games and Puzzles Journal* 10 (1989): 152–53 (British monthly).

Gorenflo, H. "Figures from Cube-Quadruples." *Praxis der Mathematik* (February 1977): 33–39.

Gossett, C. R. "A Tromino Search Problem. Problem 386." *Journal of Recreational Mathematics* 8, no. 2 (1975–76): 137–38. Solution, *JRM* 10(2): 133–34.

Graedel, Erich. "Dominoes and the Complexity of Subclasses of Logical Theories." *Annals of Pure and Applied Logic* 43, no. 1 (1989): 1–30.

Grossman, H. D. "Fun with Lattice Points: 14—A Chessboard Puzzle." *Scripta Mathematica* 14 (1948): 160.

Grünbaum, Branko, and G. C. Shephard. "Idiot-proof Tiles." *Mathematical Gazette* 75, no. 472 (June 1991): 143–47.

Comprehensive
Bibliography

————. "Patch-determined Tilings." *Mathematical Gazette* 61 (1977): 31–38.

————. "Tilings by Regular Polygons." *Mathematics Magazine* 50 (1977): 227–47; 51 (1978): 205–6.

Guttmann, A. J. "On the Number of Lattice Animals Embeddable in the Square Lattice." *Journal of Physics*. Series A, no. 15 (1982): 1987–90.

Guy, Richard K. "Problem 1122." *Crux Mathematicorum* 12–13 (1987): 50, 197–98.

Gyarfas, A., and J. Lehel. "Hypergraph Families with Bounded Edge Cover or Transversal Number." *Combinatorica* 3 (1983): 351–58.

Gyarfas, A., J. Lehel, and Zs. Tuza. "Clumsy Packing of Dominoes." *Discrete Mathematics* 71, no. 1 (1988): 33–46.

Gyoeri, Ervin. "Covering Simply Connected Regions by Rectangles." *Combinatorica* 5 (1985): 53–55.

————. "A Short Proof of the Rectilinear Art Gallery Theorem." SIAM *Journal on Algebraic and Discrete Methods* 7 (1986): 452–54.

Hajtman, Bela. "On Coverings of Generalized Checkerboards I." *Magyar Tud. Akad. Mat. Kutato Int. Köze* 7 (1962): 53–71.

Hanegraaf, Anton. "Covering a Cube by Two Hexominoes." *Cubism for Fun* (Holland; in English), no. 30 (December 1992): 27–29.

Hansson, F. Untitled. *Problemist Fairy Chess Supplement* (June-August 1935). (Later known as *Fairy Chess Review*).

Harary, Frank. "Cubical Graphs and Cubical Dimensions." *Computers and Mathematics with Applications* 15, no. 4 (1988): 271–75.

Harborth, Heiko. "Prescribed Numbers of Tiles & Tilings." *Mathematical Gazette* 61 (1977): 296–99.

Haselgrove, C. B., and Jennifer Haselgrove. "A Computer Program for Pentominoes." *Eureka* 23 (October 1960): 16–18.

Haselgrove, Jennifer. "Packing a Square with Y-pentominoes." *Journal of Recreational Mathematics* 7, no. 3 (Summer 1974): 229.

Haverman, Hans. "N-Omino Packing. Problem 929." *Journal of Recreational Mathematics* 13, no. 1 (1980–81): 57–58. Solution by B. Rosenheck, *JRM* 14(1): 69–70.

Hayes, John P. "Testing Memories for Single-Cell Pattern-Sensitive Faults." *IEEE Transactions on Computers* 29, no. 3 (1980): 249–54.

Hering, F., Ronald C. Read, and G. C. Shephard. "The Enumeration of Stack Polytopes and Simplicial Clusters." *Discrete Mathematics* 40, nos. 2–3 (1982): 203–17.

Holladay, Kenneth. "Estimating the Size of Context-free Tiling Languages." *Canadian Journal of Mathematics* 39, no. 6 (1987): 1413–33.

————. "A Partition Theory of Planar Animals." *Studies in Applied Mathematics* 63 (1980): 169–83.

Hollingsworth, Caroline. "Perplexed by 'Hexed.'" *Mathematics Teacher* 77 (October 1984): 560–62.

Jelliss, G. P. "Special Issue on Chessboard Dissections." *Chessics* 28 (Winter 1986): 137–52.

Jones, Kate. "Rhobiominoes. Problem 1961." *Journal of Recreational Mathematics* 24, no. 2 (1992): 144–46.

Josephson, B. D. "EDSAC to the Rescue." *Eureka* 24 (1961): 10–12, 32.

Judd, R. L., and M. E. Zosel. "Pentomino Alphanumerics." *Journal of Recreational Mathematics* 11, no. 3 (1978–79): 182–85.

Kadner, F. Untitled. *Problemist Fairy Chess Supplement* (February 1935). (Later known as *Fairy Chess Review*).

Kahn, Jeffry, and Michael Saks. "A Polyomino with No Stochastic Function." *Combinatorica* 4, nos. 2–3 (1984): 181–82.

Kahr, A. S., E. F. Moore, and H. Wang. "Entscheidungsproblem Reduced to the ∀∃∀ Case." *Proceedings, National Academy of Sciences USA* 48 (1962): 365–77.

Katona, G., and D. Szász. "Matching Problems." *Journal of Combinatorial Theory*. Series B, 10, no. 1 (1971): 60–92.

Keith, M. "A Pentomino Conjecture. Problem 391." *Journal of Recreational Mathematics* 8, no. 2 (1975–76): 140. Solutions, *JRM* 10(2): 136–39.

——. "A Pentomino Query. Problem 426." *Journal of Recreational Mathematics* 8, no. 3 (1975–76): 231. Solution by P. J. Torbijn, *JRM* 10(3): 219.

Kelly, J. B. "Polynomials and Polyominoes." *American Mathematical Monthly* 73 (1966): 464–71.

Kim, Dongsu. "The Number of Convex Polyominoes with Given Perimeter." *Discrete Mathematics* 70, no. 1 (1988): 47–51.

Klamkin, M. S., and Andy C. F. Liu. "Polyominoes on the Infinite Checkerboard." *Journal of Combinatorial Theory*. Series A, 28, no. 1 (1980): 7–16.

Klarner, David A. "Brick-packing Puzzles." *Journal of Recreational Mathematics* 6, no. 2 (1973): 112–17.

——. "Cell Growth Problems." *Canadian Journal of Mathematics* 19 (1967): 851–63.

——. "Letter to the Editor." *Journal of Recreational Mathematics* 3, no. 4 (October 1970): 258.

——. "Packing a Rectangle with Congruent N-ominoes." *Journal of Combinatorial Theory* 7 (1969): 107–15.

——. "A Packing Theory." *Journal of Combinatorial Theory*. Series A, 8 (1970): 273–78.

——. "A Search for N-pentacube Prime Boxes." *Journal of Recreational Mathematics* 12, no. 4 (1979–80): 252–57.

——. "Some Results Concerning Polyominoes." *Fibonacci Quarterly* 3 (1965): 9–20.

Klarner, David A., and F. Goebel. "Packing Boxes with Congruent Figures." *Indagationes Mathematicae* (Amsterdam) 31 (1969): 465–72. MR 80a#05067.

Comprehensive
Bibliography

Klarner, D. A., and M. L. J. Hautus. "Uniformly Coloured Stained Glass Windows." *London Mathematical Society Proceedings*. Series 3, part 4 (1971): 613–28.

Klarner, David A., and R. Rivest. "A Procedure for Improving the Upper Bound for the Number of N-ominoes." *Canadian Journal of Mathematics* 25 (1973): 585–602.

———. "Asymptotic Bounds on the Number of Convex n-Ominoes," *Discrete Mathematics* 8, no. 1 (1974): 31–40.

Klarner, David A., and D. Szász. "Matching Problems." *Journal of Combinatorial Theory*. Series B, 10 (1971): 60–92.

Knop, J. V., K. Szymanski, Z. Jericevic, and N. Trinajstic. "On the Total Number of Polyhexes." *Match* 16 (1984): 119–34.

Kotlyar, B. D. "Packings of Parallelotopes and Certain Other Sets." Akademiya Nauk SSSR. Sibirskoe Otdelenie. *Sibirskii Matematicheskii Zhurnal* 25, no. 2 (1984): 222–25. MR 85j#52021.

Kramer, Earl S. "Tiling Rectangles with T and C Pentominoes." *Journal of Recreational Mathematics* 16 (1983): 102–13.

Kramer, Earl S., and Frits Goebel. "Tiling Rectangles with Pairs of Pentominoes." *Journal of Recreational Mathematics* 16, no. 3 (1983–84): 198–206.

Laatsch, Richard G. "Rectangles from Mixed Polyomino Sets." *Journal of Recreational Mathematics* 13, no. 3 (1980–81): 183–87.

Langford, C. Dudley. "Note 1464: Uses of a Geometric Puzzle." *Mathematical Gazette* 24, no. 260 (July 1940): 209–11.

———. "Note 2793: A Conundrum for Form VI." *Mathematical Gazette* 42, no. 342 (December 1958): 287.

———. "Note 2864: A Chess-board Puzzle." *Mathematical Gazette* 43, no. 345 (October 1959): 200.

Lee, Sin-Min, and Rudy Tanoto. "Three Classes of Diameter Edge-invariant Graphs." *Commentationes Mathematicae, Univ. Karolova* 28 (1987): 227–32.

Levine, Jack. "Note on the Number of Pairs of Non-Intersecting Routes." *Scripta Mathematica* 24, no. 4 (1959): 335–38.

Lin, K'e Ying, and S. J. Chang. "Rigorous Results for the Number of Convex Polygons on the Square and Honeycomb Lattices." *Journal of Physics*. Series A, Mathematical and General 21, no. 11 (1988): 2635–42.

Liu, Andy C. F. "Pentomino Problems." *Journal of Recreational Mathematics* 15, no. 1 (1982–83): 8–13.

Loyd, S. "An Ancient Puzzle (No. 18)" *Tit-Bits* 31 (February 13, March 6, 1897): Origin of a Famous Puzzle, 363, 419.

———. "The Mitre Puzzle (No. 19)" *Tit-Bits* 31 (February 13, March 6, 1897): Origin of a Famous Puzzle, 363, 419.

Lunnon, W. F. "Counting Multidimensional Polyominoes." *Computer Journal* 18 (1975): 366–67.

Macdonald, D., and Y. Gürsel. "Solving Soma Cube and Polyomino Puzzles Using a Microcomputer." *BYTE* 4, no. 11 (November 1979): 26–50.

Mackinnon, Nick. "Some Thoughts on Polyomino Tiles." *Mathematical Gazette* 74, no. 467 (March 1990): 31–33.

Madachy, Joseph S. "Pentominoes—Some Solved and Unsolved Problems." *Journal of Recreational Mathematics* 2, no. 3 (July 1969): 181–88.

————. "Recreational Mathematics." *Fibonacci Quarterly* 6, no. 2 (1968): 162–66.

Madras, N., C. E. Soteros, and S. G. Whittington. "Statistics of Lattice Animals." *Journal of Physics*. Series A, 21, no. 24 (1988): 4617–35.

Maloney, John P. "A Non-Graph Theory Approach to Ulam's Conjecture." *Portugaliae Mathematica* 36 (1977): 1–6.

Marlow, T. W. "Grid Dissections." *Chessics* 23 (Autumn 1985): 78–79.

Martin, George E. "Polytaxic Polygons." *Structural Topology. Topologie Structurale,* no. 12 (1986): 5–10.

Masalski, W. J. "Polycubes." *Mathematics Teacher* 70, no. 1 (January 1977): 46–50.

Mayer, Jean. "A Pentomino Problem." *Journal of Recreational Mathematics* 6, no. 2 (Spring 1973): 105–8.

Meeus, Jean. "The Smallest U-N Square." *Journal of Recreational Mathematics* 18, no. 1 (1985–86): 8.

————. "Some Polyomino and Polyiamond Problems." *Journal of Recreational Mathematics* 6, no. 3 (Summer 1973): 215–20.

————. "Tetracubes." *Journal of Recreational Mathematics* 6, no. 4 (1973): 257–65.

————. "Tiling Rectangles with Pairs of Pentominoes." *Journal of Recreational Mathematics* 18, no. 1 (1985–86): 49–51.

Meeus, Jean, and Pieter J. Torbijn. "The 30- and 36-Problems with Pentominoes and Hexiamonds." *Journal of Recreational Mathematics* 10, no. 4 (1977–78): 260–66.

Mertens, S. "Lattice Animals: A Fast Enumeration Algorithm and New Perimeter Polynomials." *Journal of Statistical Physics* 58, nos. 5–6 (1990): 1095–1108.

Miller, J.C.P. "Pentominoes." *Eureka* 23 (October 1960): 13–18.

Motoyama, Akiko, and Haruo Hosoya. "Tables of the King and Domino Polynomials for Polyominoes." *Natural Science Reports* Ochanomizu Univ. 27, no. 2 (1976): 107–23.

Myers, Basil R. "Enumeration of Tours in Hamiltonian Rectangular Lattice Graphs." *Mathematics Magazine* 54 (1981): 19–23.

Nelson, H. L. "Hexomino Packing. Problem 1064." *Journal of Recreational Mathematics* 14, no. 2 (1981–82): 138. Solution by R. I. Hess, *JRM* 15(2): 145.

Niemann, J. Untitled. *Fairy Chess Review* (June 1938, December 1939).

————. Untitled. *Fairy Chess Review* 7 (1948): 8.

O'Beirne, T. H. "For Boys, Men and Heroes." *New Scientist,* no. 266 (December 21, 1961): Puzzles and Paradoxes, 751–52.

————. "Pell's Equation in Two Popular Problems." *New Scientist,* no. 258 (October 26, 1961): Puzzles and Paradoxes, 260–61.

Comprehensive
Bibliography

————. "Pentominoes and Hexiamonds." *New Scientist*, no. 259 (November 2, 1961): Puzzles and Paradoxes, 316–17.

————. "Some Hexiamond Solutions and an Introduction to a Set of 25 Remarkable Points." *New Scientist*, no. 260 (November 9, 1961): Puzzles and Paradoxes, 379–80.

————. "Some Tetrabolical Difficulties." *New Scientist,* no. 270 (January 18, 1962): Puzzles and Paradoxes, 158–59.

Ohno, Yoshio. "Pentomino Packing II." *Journal of Recreational Mathematics* 15, no. 2 (1982–83): 143. Solution by Wade E. Philpott, *JRM* 16(2): 149–50.

Olivastro, Dominic. "Fat Dominoes." *The Sciences* (July-August 1992): 53–55.

Patton, Robert L. "Pentomino Farms." *Journal of Recreational Mathematics* 1, no. 4 (October 1968): Letter to the Editor, 234–35.

Penaud, J.-G. "Arbres et Animaux." Ph.D. diss., Université de Bordeau I, May 1990.

Penner, Sidney. "Tiling a Checkerboard with Dominoes. Problem E2508." *American Mathematical Monthly* 83 (March 1976): 199–200.

Peterson, Ivars. "Pieces of a Polyomino Puzzle." *Science News* 132, no. 20 (November 14, 1987): 310.

Philpott, Wade E. "Polyomino and Polyiamond Problems." *Journal of Recreational Mathematics* 10, nos. 1, 2 (1977–78): 2–14, 98–104.

————. "Polyominoes of Orders One Through Seven." *Journal of Recreational Mathematics* 13, no. 1 (1980–81): 58.

Piazza, Barry L., and Richard D. Ringeisen. "A Combinatorial Analysis of Preston's Dodecamino Table." *Journal of Combinatorics, Information & System Sciences* 12, nos. 1–2 (1987): 66–74.

Povah, Maurice J. "Letter." *Mathematical Gazette* 45, no. 354 (December 1961): 342.

Prentice, Allan, Pieter J. Torbijn, and Dario Uri. "Solid Pentomino Helix. Parts 1, 2." *Journal of Recreational Mathematics* 21, no. 2 (Summer 1989): 150–52.

Rands, B.M.I. and D.J.A. Welsh. "Animals, Trees and Renewal Sequences." IMA. *Journal of Applied Mathematics* 27, no. 1 (1981): 1–17.

Rawsthorne, Daniel A. "Tiling Complexity of Small N-ominoes [N < 10]." *Discrete Mathematics* 70, no. 1 (1988): 71–75.

Read, Ronald C. "Contributions to the Cell Growth Problem." *Canadian Journal of Mathematics* 14, no. 1 (1962): 1–20.

————. "The Dimer Problem for Narrow Rectangular Arrays: A Unified Method of Solution, and Some Extensions." *Aequationes Mathematicae* 24 (1982): 47–65.

————. "On General Dissections of a Polygon." *Aequationes Mathematicae* 18 (1978): 370–88.

Redelmeier, D. Hugh. "Counting Polyominoes: Yet Another Attack." *Discrete Mathematics* 36, no. 2 (1981): 191–203.

Reeve, J. E., and J. A. Tyrrell. "Maestro Puzzles." *Mathematical Gazette* 45 (October 1961): 97–99.

Rensburg, E. J. Janse van. "The Topology of Interfaces." *Journal of Physics*. Series A, Mathematical and General 23, no. 24 (1990): 5879–95.

Rensburg, E. J. Janse van, and S. G. Whittington. "Self-avoiding Surfaces." *Journal of Physics*. Series A, Mathematical and General 22, no. 22 (1989): 4939–58.

Riele, Herman, J. J. te, and D. T. Winter. "The Tetrahexes Puzzle." *CWI Newsletter* (Amsterdam) 10 (March 1986): 33–39.

Roothart, Chris. "Square-Free Polytans." *Cubism for Fun* (Holland; in English), no. 30 (December 1992): 16–17.

Sachs, Horst. "Perfect Matchings in Hexagonal Systems." *Combinatorica* 4 (1984): 89–99.

Sampson, J. R., and S. C. Trofanenko. "Aspects of Shape." INFOR. *Canadian Journal of Operational Research and Information Processing* 17, no. 2 (1979): 138–50.

Sands, B. "The Gunport Problem." *Mathematics Magazine* 44 (1932): 193–96.

Schattschneider, Doris. "Will It Tile? Try the Conway Criterion!" *Mathematics Magazine* 53, no. 4 (1980): 224–33.

Scherer, Karl. "Heptomino Tessellations. Problem 1045." *Journal of Recreational Mathematics* 14, no. 1 (1981–82): 64. Solution by Scherer, *JRM* 21(3): 221–23; Solution by Dahlke, *JRM* 22(1): 68–69.

———. "Minimal Fault-free Rectangles Packed with I_n-Polyominoes." *Journal of Recreational Mathematics* 13, no. 1 (1980–81): 4–6.

———. "Some New Results on Y-pentominoes." *Journal of Recreational Mathematics* 12, no. 3 (1979–80): 201–4.

———. "Tessellation II. Problem 982." *Journal of Recreational Mathematics* 13, no. 3 (1980–81): 218. Solution by J. Meeus, *JRM* 14(3): 224–25.

———. "Tessellation II, Again. Problem 1665." *Journal of Recreational Mathematics* 20, no. 3 (1990): 228. Solution by proposer; other solutions by S. Higgins, J. Verbakel, J. Spain, *JRM* 22(3): 229–32.

Shao, Jiayu. "Matrices Permutation Equivalent to Primitive Matrices." *Linear Algebra and Its Applications* 65 (1985): 225–47.

Shearer, James B. "Barring Hexominoes from the Infinite Checkerboard." *Studies in Applied Mathematics* 67, no. 3 (1982): 243–55.

———. "Barring the Z Pentomino from the Infinite Checkerboard." *Studies in Applied Mathematics* 67, no. 1 (1982): 73–77.

———. "A Class of Perfect Graphs." *SIAM Journal on Algebraic and Discrete Methods* 3, no. 3 (1982): 281–84.

Sibson, R. "Note 1485: Comments on Note 1464." *Mathematical Gazette* 24, no. 262 (December 1940): 343.

Singmaster, David. "Covering Deleted Chessboards with Dominoes." *Mathematics Magazine* 48 (1975): 59–66.

Comprehensive
Bibliography

Spearman, John. "Close Packing." *Mathematical Gazette* 70 (June 1986): 89–91.

Spira, Robert. "Problem E1983." *American Mathematical Monthly* 74, no. 4 (April 1967): 439. Solution by Dennis Gannon, *AMM* 75(7): 785–86.

Stanley, Richard P. "On Dimer Coverings of Rectangles of Fixed Width." *Discrete Applied Mathematics* 12, no. 1 (1985): 81–87.

Stead, W. "Dissection." *Fairy Chess Review* 9, no. 1 (December 1954): 2–4.

Stewart, Ian, and A. Wormstein. "Polyominoes of Order 3 Do Not Exist." *Journal of Combinatorial Theory*. Series A, 61, no. 1 (1992): 130–36.

"Sūgei Pazuru" (Mathematical Art of Puzzles). Published by Sūgei Pazuru Aikokai (Mathematical Art of Puzzles Club), Nagoya, Japan, no. 153 (November-December 1986): 64–81.

Szabó, S. "A Bound of k for Tiling by (k,n) Crosses and Semicrosses." *Acta Mathematica Hungarica* 44, nos. 1–2 (1984): 97–99.

Takefuji, Yoshiyasu, and Kuo-Chun Lee. "A Parallel Algorithm for Tiling Problems." IEEE *Transactions on Neural Networks* 1, no. 1 (March 1990): 143–45.

"31: Polyominoes." *QARCH* 1, no. 8 (June 1984): 11–13. (An occasional publication of The Archimedeans, Cambridge).

Torbijn, Pieter J. "Black Holes." *Cubism for Fun* (Holland; in English), no. 29 (September 1992): 14–15.

———. "Covering a Cube with 36 Hexominoes." *Cubism for Fun* (Holland; in English), no. 30 (December 1992): 33.

———. "Cubic Hexomino Cubes." *Cubism for Fun* (Holland; in English), no. 30 (December 1992): 18–19.

———. "General Hexomino Cubes." *Cubism for Fun* (Holland; in English), no. 30 (December 1992): 21–26.

———. "The Jagged Square. Problem 1791." *Journal of Recreational Mathematics* 22, no. 2 (1990): 141–42. Solutions by proposer and by D. Wilms, C. Ashbacher, R. Hess, B. Barwell, M. Keller, M. E. Larsen, *JRM* 23(2): 145–47, 199.

———. "Polyiamonds." *Journal of Recreational Mathematics* 2, no. 4 (October 1969): 216–27.

———. "A Solid Pentomino Allen Rocket." *Journal of Recreational Mathematics* 22, no. 3 (Fall 1990): 234–35.

Torbijn, Pieter J., and Jean Meeus. "Hexominoes in Rectangles." *Journal of Recreational Mathematics* 19, no. 4 (1987): 254–60.

———. "Hexomino Cubes." *Cubism for Fun* (Holland; in English), no. 28 (April 1992): 9.

Treep, Anneke. "The Narrow Passage Problem. Problem 1739." *Journal of Recreational Mathematics* 21, no. 3 (1989): 220. Solution by proposer, Friend H. Kierstead, Jr., and P. J. Torbijn, *JRM* 22(3): 237–38.

Turner, John Christopher. "On Folyominoes and Feudominoes." *Fibonacci Quarterly* 26, no. 3 (1988): 205–18.

"Two Dissection Problems: 2." *Eureka* 13 (1950): 6; 14 (1951): 23.

Underwood, Val. "Polyominoes." *Mathematics Teacher* 41 (1967): 54–57.

"Unequally Modified Checkerboard. Problem 170." *Second-Year College Mathematics Journal* 17 (November 1981): 345–47.

Vejmola, Stan. "A Pentomino Area Problem. Problem 1660." *Journal of Recreational Mathematics* 20, no. 3 (1988): 226–27. Solution by Brian Barwell, *JRM* 21(3): 233; solution by P. J. Torbijn and Jan Verbakel, *JRM* 22(3): 229.

———. "Variations on Pentomino Farms. Problem 1661." *Journal of Recreational Mathematics* 20, no. 3 (1988): 227. Solutions by proposer and by P. J. Torbijn, *JRM* 21(3): 233–34.

Vejmola, Stan, Brian Barwell, and Pieter J. Torbijn. "A Pentomino Area Problem." *Journal of Recreational Mathematics* 21, no. 3 (Fall 1989): 233.

Vejmola, Stan, Pieter J. Torbijn, and Jan Verbakel. "A Pentomino Area Problem." *Journal of Recreational Mathematics* 22, no. 3 (Fall 1990): 229.

Verbakel, Jan. "The F-Pentacube Problem." *Journal of Recreational Mathematics* 5, no. 1 (January 1972): 20–21.

Wagner, N. R. "Constructions with Pentacubes. Parts 1, 2." *Journal of Recreational Mathematics* 5–6, nos. 4, 3 (1972, 1973): 266–68, 211–14.

Wagon, Stan. "Fourteen Proofs of a Result about Tiling a Rectangle." *American Mathematical Monthly* 94 (1987): 601–17.

Walkup, D. W. "Covering a Rectangle with T-tetrominoes." *American Mathematical Monthly* 72, no. 9 (November 1965): 986–88.

Wells, David. "Dissecting N-ominoes to Squares." *Games and Puzzles Journal* (October-November 1974): 38 (British monthly).

Wijshoff, H.A.G., and Jan van Leeuwen. "Arbitrary Versus Periodic Storage Schemes and Tessellations of the Plane Using One Type of Polyomino." *Information and Control* 62, no. 1 (1984): 1–25.

Journal Articles

Name Index